SpringerBriefs in Statistics

More information about this series at http://www.springer.com/series/8921

Rosa Arboretti · Arne Bathke · Stefano Bonnini
Paolo Bordignon · Eleonora Carrozzo
Livio Corain · Luigi Salmaso

Parametric and Nonparametric Statistics for Sample Surveys and Customer Satisfaction Data

 Springer

Rosa Arboretti
Department of Civil and Environmental
Engineering
University of Padova
Padova, Italy

Stefano Bonnini
Department of Economics and Management
University of Ferrara
Ferrara, Italy

Eleonora Carrozzo
Department of Management and Engineering
University of Padova
Padova, Italy

Luigi Salmaso
Department of Management and Engineering
University of Padova
Padova, Italy

Arne Bathke
Natural Science
University of Salzburg
Salzburg, Austria

Paolo Bordignon
Department of Management and Engineering
University of Padova
Padova, Italy

Livio Corain
Department of Management and Engineering
University of Padova
Padova, Italy

ISSN 2191-544X   ISSN 2191-5458   (electronic)
SpringerBriefs in Statistics
ISBN 978-3-319-91739-9   ISBN 978-3-319-91740-5   (eBook)
https://doi.org/10.1007/978-3-319-91740-5

Library of Congress Control Number: 2018943287

Printed on acid-free paper

This Springer imprint is published by the registered company Springer International Publishing AG part of Springer Nature.
The registered company address is: Gewerbestrasse 11, 6330 Cham, Switzerland

Preface

This book deals with problems related to the evaluation of customer satisfaction in very different contexts and in many different ways. Analyzing satisfaction is not an easy issue since it represents a complex phenomenon which is not directly measurable. Often satisfaction about a product or service is investigated through suitable surveys which try to capture the satisfaction about several partial aspects which characterize the perceived quality of that product or service.

In this book we present a series of statistical techniques adopted to analyze data from real situations where customer satisfaction surveys were performed. The aim is to give a simple guide of the variety of analysis that can be performed when analyzing data from surveys.

Experiencing satisfaction when customers buy products or services is an index of how a company is operating. Are goods or services appreciated by customers? Are customers satisfied with goods or services of a specific company?

How they respond to similar questions is a crucial point in order to evaluate and analyze the answers. For this purpose preference evaluation methods are good candidates to understand how customers react to the evaluated items. A promising and theory-based method is called CUB model. In the first chapter of this book CUB model has been adopted in order to evaluate two latent variables, feelings and uncertainty, that are supposed to be involved in the choice process of an item. The application field refers to a genuine study conducted in Italy and in Austria about the satisfaction level of customers about food packaging at the grocery store.

The second chapter deals with the concept of heterogeneity in satisfaction. Identifying customer groups characterized by "within homogeneity" and "between heterogeneity" could be a useful starting point of market segmentation. In this chapter the main heterogeneity indices are introduced and testing methods for comparing the satisfaction heterogeneities of two or more customer populations are described also with different practical examples.

In the field of satisfaction assessment it is quite common that the final objective is obtaining an appropriate ordering of different products or services under comparison. From a statistical point of view the issue of ranking several populations from the best to the worse on the basis of one or more aspects of interest is not so easy. In the third chapter of this book different examples of contexts where the problem of ranking occurs are described and a nonparametric inferential approach is presented with application to the field of food sensory analysis.

Another way to assess satisfaction is represented by the so-called composite indicators which aggregate different dimensions of satisfaction into a single overall indicator. How to suitably compute such indicator is the topic of Chap. 4. In this chapter the construction of a composite indicator is discussed in general and a nonparametric composite indicator which includes different benchmarks of satisfaction is developed. The properties of the proposed indicator are shown by analyzing data from a university students' satisfaction survey.

Finally Chap. 5 describes some rank-based procedures for analyzing surveys data with the help of a useful R package.

Padova, Italy   Rosa Arboretti
Salzburg, Austria   Arne Bathke
Ferrara, Italy   Stefano Bonnini
Padova, Italy   Paolo Bordignon
Padova, Italy   Eleonora Carrozzo
Padova, Italy   Livio Corain
Padova, Italy   Luigi Salmaso
March 2018

Acknowledgments

The work was also partially supported by the University of Ferrara, which funded the FIR (Research Incentive Fund) project "Advanced Statistical Methods for Data Analysis in Complex Problems."

Contents

Chapter 1
The CUB Models

The CUB model [12], where CUB stands for Combination of a discrete Uniform and a shifted Binomial distributions assumes the involvement of two latent variables during an evaluation process, that have been called *feeling* and *uncertainty*. In order to justify the names for latent variables, consider the way you choose an evaluation grade from a set of 9. The final choice reflects your feeling about the evaluated item, your past experience, your knowledge about it, and so on. On the other hand, there are some aspects concern with a basic uncertainty about the evaluated item, for example you are asked to deal with it for the first time and you don't know what grade to choose, maybe the task is too difficult or the task is annoying you. These two main components are supposed to move your final choice and they are supposed to follow respectively a shifted Binomial distribution and a Uniform distribution [12, 18]. The CUB models were first described as a suitable method for preference evaluation [12]. There are two main approaches for evaluating preferences i.e. stated and revealed preferences (see e.g. [1, 22]) and CUB models have been considered a stated preference approach [18]. The probabilistic structure described by D'Elia and Piccolo [12] considers the psychological process of evaluating a specific item in a survey, where subjects are usually asked to evaluate some items (products, services, etc.) by means of ranking or rating scales. The second one is usually the most preferred by respondents. One of the first applications of CUB models was described by Piccolo and D'Elia [25] where models were applied to a large data-set on preferences about smoked salmons. Subjects were asked to evaluate five brands of smoked salmons on a 9-point scale. In a first step, the authors estimated the two latent variables, *feeling* and *uncertainty*, subsequently they introduced subject's and object's covariates into the model in order to link some relevant information to the latent variables. The CUB model with covariates has been extensively applied in many studies after their introduction [5, 6]. CUB models have also been applied for

© The Author(s), under exclusive licence to Springer International Publishing AG, part of Springer Nature 2018
R. Arboretti et al., *Parametric and Nonparametric Statistics for Sample Surveys and Customer Satisfaction Data*, SpringerBriefs in Statistics, https://doi.org/10.1007/978-3-319-91740-5_1

customer satisfaction investigations and they have been proven to help at dealing with marketing questions. For instance, in a study on wine consumers CUB model with covariates have been applied in order to identify which wine characteristics are more relevant for Italian wine consumers [8]. For other examples of applications to marketing issues see [20, 21]. The following section will describe the CUB model structure and some extensions.

1.1 Description of CUB Models Structure

The CUB model is a mixture model that aims at evaluating two latent variables, *feeling* and *uncertainty*, that are supposed to be involved in the choice process of an item. The process of selecting a grade among m can be represented by the mixture of two components: the liking—disliking towards the evaluating item (products or services) and an inherent uncertainty that belongs to any human choice [12]. The two components can be described by a mixture model of two random variables, *feeling* and *uncertainty*, with a shifted Binomial and a Uniform distribution respectively [12, 25]. When respondents are involved in the choice process of an evaluation grade, they synthesize what they feel so that it has been adopted the wide term *feeling* to cover the several psychological aspects surrounding the final choice, such as motivation, awareness, attraction, knowledge, etc. A shifted Binomial random variable is supposed to simulate the mechanism of selecting an evaluation grade by means of paired comparisons among the grades [9]. Let us consider the scale $y = 1, \ldots, m$. In the choice of a grade y, you show to prefer that point over the lower and higher ones because they are not matching what you feel. Let p and $1 - p$ be the probability of rejecting a point respectively because it is too low and too high compared to y, so the probability of choosing a grade y can be described by a shifted Binomial distribution as follows [19]:

$$Pr(Y = y) = \binom{m-1}{y-1} p^{y-1}(1-p)^{my} \tag{1.1}$$

The *uncertainty* is considered a "gray zone", where the choice of respondents is affected by several aspects, e.g. limited time and knowledge, inherent aptitude to give fake answers, etc. The maximum expression of these aspects describes a pattern of responses where each choice has the same probability to occur. A probability distribution that follows such a pattern is the Uniform discrete distribution $U_y(m) = \frac{1}{m}, y = 1, \ldots, m$. It should be noted that while *uncertainty* expresses the inherent fuzziness of any human choice, randomness is linked to statistical aspects such as measurement errors, sample methods and so on [18]. The components *feeling* and *uncertainty* are combined in a mixture model whose probability function of the random variable Y is as follows [7, 11, 12, 17, 18]:

$$Pr(Y = y) = \pi \binom{m-1}{y-1}(1-\xi)^{y-1}\xi^{m-y} + \frac{(1-\pi)}{m} \tag{1.2}$$

$y = 1, 2, \ldots, m$, where Y varies from 1 to m, $\xi \in [0,1]$, $\pi \in (0,1]$. Iannario [16] shows that the model is identified for $m > 3$. The parameter vector $\theta = (\pi, \xi)$ belongs to the parameter space $\Omega(\theta) = \{(\pi, \xi) : 0 < \pi \le 1, 0 \le \xi \le 1\}$. This is a very flexible distribution because it can assume many different shapes considering we have only two parameters, π and ξ [12, 23].

The parameter π is an estimate of *uncertainty* and $(1 - \pi)$ is considered a measure of uncertainty with $(1 - \pi)/m$ the uncertainty shared by response categories. Parameter ξ estimates the *feeling* component that affects a choice process, whose interpretation depends on scale coding. Let us consider a 9-point scale with $y = 1$ as minimum, then an estimate of ξ close to 1 indicates a *feeling* component very close to minimum. In this case, the distribution of observed values has a mode close to or at the point $y = 1$, so that $(1 - \xi)$ can be considered a measure of *feeling*. In this case, high values of $(1 - \xi)$, i.e. values close to 1, indicate a mode close to $y = 9$ which means a high feeling/liking. In order to show the flexibility of the model, a graphic (see Fig. 1.1) is aimed to show the distribution shapes of CUB model varying π and ξ parameter values. When $\xi = 0.5$ the distribution is symmetric with mode on the central value. As $(1 - \xi)$ parameter increases from 0 to 1, the mode values increase showing that $(1 - \xi)$ can be considered a direct measure of liking [18]. About *uncertainty*, when π increases the distribution assumes a bell shape and values on y axis increase rapidly. A pattern like this would mean a low *uncertainty* among subjects' responses.

Fig. 1.1 The distribution shapes of CUB model varying π and ξ parameter values

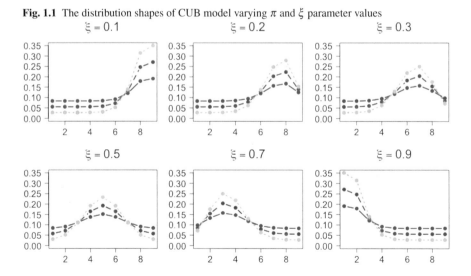

D'Elia [10] and Piccolo [24] developed an E-M algorithm for the maximum likelihood estimate of parameters π and ξ and they present a detailed description of the E-M algorithm.

1.1.1 Model Extensions

Since their introduction, CUB models have been supplied with several extensions. The main scope of extensions was to let CUB models take into considerations some aspects affecting the choice process, for instance the introduction of covariates into the model. D'Elia [10] and Piccolo [24] provide a formal description of CUB model with covariates. Information about subjects or objects can be introduced into the model and linked to *uncertainty* and *feeling* by means of two logistic functions. Consider the following specification of CUB model:

$$Pr(Y_i = y_i) = \pi_i \binom{m-1}{y_i-1}(1-\xi_i)^{y_i-1}\xi_i^{m-y_i} + \frac{(1-\pi_i)}{m} \tag{1.3}$$

with $y = 1, 2, \ldots, m$. Parameters π_i ad ξ_i are explained by two covariate vectors, $X_i = (1, x_{i1}, \ldots, x_{ip})$ and $W_i = (1, w_{i1}, \ldots, w_{iq})$, that are linked to parameters by the following relations:

$$\pi_i = \frac{1}{1+\exp^{-\beta_0-\beta_1 x_{i1}-\ldots-\beta_p x_{ip}}} = \frac{1}{1+\exp^{-\beta_0-x_i\beta}} \tag{1.4}$$

and

$$\xi_i = \frac{1}{1+\exp^{-\gamma_0-\beta_1 w_{i1}-\ldots-\gamma_q w_{iq}}} = \frac{1}{1+\exp^{-\gamma_0-w_i\gamma}}. \tag{1.5}$$

The CUB(p,q) model indicates p covariates for π and q covariates for ξ. By this way, parameters can be linked to subjects' covariates (gender, age, income, etc.) or objects' covariates (some characteristic of the item), showing which information affects *feeling* ad *uncertainty*. The CUB model with covariates is an effective approach to identify which objects' or subjects' characteristics are relevant in the choice process. In particular, the approach can delineate clusters based on significant covariates, which outline groups of subjects that behave differently in terms of *feeling* and *uncertainty*. For example, Piccolo and D'Elia [25] applying CUB models with covariates identified a relation between gender and age in the likeness of smoked salmons. Moreover they showed how sensory perception changes with respect to different chemical characteristics of smoked salmon. Furthermore Iannario and Piccolo [18] applied CUB models to investigate the link between satisfaction and some relevant characteristics of a product. Iannario [15] introduced the concept of *shelter choice* that has been supposed to be present in surveys with atypical patterns of observed frequencies. An atypical pattern can be the results of an over-selection of a specific grade because of unwillingness to respond or privacy reasons.

Because of a difficult choice, subjects sometimes tend to simplify a response by selecting the same grade of judgment. For instance, lazy subjects might prefer central points of a rating scale in order to simplify a more demanding response. When they are dealing with categories like satisfy, very satisfy, extremely satisfy, sometimes the first positive category, i.e. satisfy, is chosen, so as to avoid a more elaborated response. When an observed frequency is higher than an estimated one by CUB model or when there are reasons to say that a certain point, let say $y = c$, could be a shelter choice, should be taken into consideration an adequate extension of the CUB model. Iannario [18] developed the following extension of the CUB model in order to estimate the *shelter effect*:

$$Pr(Y = y) = \pi_1 \binom{m-1}{y-1}(1-\xi)^{y-1}\xi^{m-y} + \frac{\pi_2}{m} + (1-\pi_1+\pi_2)D_y^{(c)}. \quad (1.6)$$

The CUB model with shelter choice describes the probability distribution of a random variable Y, with $\theta = (\pi_1, \pi_2, \xi)$ the parameter vector belonging to the parameter space defined as $\Omega(\theta) = \{(\pi_1, \pi_2, \xi) : \pi_1 > 0, \pi_2 \geq 0, \pi_1 + \pi_2 \leq 1, 0 \leq \xi \leq 1\}$. $D_y^{(c)}$ is a degenerate random variable with $D_y^{(c)} = 1$ when $y = c$ and $D_y^{(c)} = 0$ elsewhere. The equivalence $\delta = 1 - \pi_1 - \pi_2$ quantity the shelter effect at $Y = c$. The model can be formulated again considering the parameter vector $\Theta = (\pi, \xi, \delta)$ as follows:

$$Pr(Y = y) = (1-\delta)[\pi b_y(\xi) + (1-\pi)U_y] + \delta D_y^{(c)} \quad (1.7)$$

where b_y is the shifted Binomial distribution and U_y is the discrete Uniform distribution. Considering the two formulations of the CUB model with shelter effect, the following relation among parameters can be determined:

$$\begin{cases} \pi_1 = (1-\delta)\pi \\ \pi_2 = (1-\delta)(1-\pi) \end{cases} \Longleftrightarrow \begin{cases} \pi = \frac{\pi_1}{\pi_1+\pi_2} \\ \delta = 1 - \pi_1 - \pi_2 \end{cases}$$

A detailed explanation of the L-M algorithm for parameter estimates of CUB model with shelter effect is reported in Iannario [15]. CUB models estimate a probability distribution related to parameters that are supposed to measure latent variables. In order to evaluate the goodness of the model, estimated distribution and observed distribution could be compared. When an estimated distribution does not fit very well to the observed one and when variability is very high, e.g. presence of more than one mode or high variance, a CUB model extension can be applied. For instance, a CUB model could identify clusters of people describing very different observed distributions, or there could be a shelter effect. A third solution to improve bad fitting is to measure *overdispersion*. The concept of *overdispersion* has been introduced by Iannario [17] suggesting a high variability among subjects' *feeling* component. Overdispersion affects the *feeling* latent variable, the way subjects react to evaluated items. In other words, overdispersion suggests an inter-personal way to make choices that is very different among respondents. In order to modeled overdispersion, Iannario [17] introduced a Beta-binomial distribution in the CUB model. The model with a Beta-binomial distribution has been called CUBE model (*Combination* of a *U*niform and a shifted *BE*ta-binomial).

The model

$$Pr(Y = y) = \pi \cdot be(\xi, \phi) + (1 - \pi)\frac{1}{m}, y = 1, \ldots, m \tag{1.8}$$

represents the probability distribution of a random variable Y with the parameter vector $\theta = (\pi, \xi, \phi)$ belonging to the parameter space $\Omega(\theta) = (\pi, \xi, \phi) : 0 < \pi \leq 1$, $0 \leq \xi \leq 1, 0 \leq \phi < \infty$, and with the parameter $\phi \neq 0$ indicating an *overdispersion* effect. In CUBE model, a relevant part is represented by a Beta-binomial distribution $be(\xi, \phi)$ whose derivation is explained in [17]. The Beta-binomial distribution is as follows,

$$Pr(Y = y) = be(\xi, \phi) =$$

$$= \binom{m-1}{y-1} x \frac{\prod_{k=1}^{y} [1 - \xi + \phi(k-1)] \prod_{k=1}^{m-y+1} [\xi + \phi(k-1)]}{[1 - \xi + \phi(y-1)][\xi + \phi(m-y)] \prod_{k=1}^{m-1} [1 + \phi(k-1)]} \tag{1.9}$$

$y = 1, \ldots, m$, where parameters ξ and ϕ are linked to *feeling* and overdispersion respectively.

Let us consider the two central moments of first order of a Beta-binomial random variable $E(Y) = \xi + m(1 - \xi)$ and $Var(Y) = (m-1)\xi(1-\xi)(1 + \frac{(m-2)\phi}{1+\phi})$ variance increases of an amount depending on parameter $\phi > 0$, that is linked to overdispersion. The CUB models described so far give an overview of their great power in explaining more than one behaviors related to a choice process. We stressed particularly on conceptual meaning of CUB models, notwithstanding we gave specific references for a more detailed description of statistical derivation of models.

1.2 Fitting Measures

Models are usually considered useful tools for studying phenomena, but are they adequately representing the object observed? In order to evaluate if a model is useful, we need to have a measure of model fitting to observed data. In fact, all models are wrongs but some are useful [4]. CUB models estimate probabilities given a parameter vector θ. The estimated probabilities $p_y(Y = y|\theta)$ should be very close to observed probabilities f_y when the model is good. The following index of fitting, called *normalized dissimilarity index*, is a measure of distance between estimated and observed frequencies:

$$Diss = 0.5 \sum_{y=1}^{m} |f_y - p_y(\theta)|. \tag{1.10}$$

The normalized dissimilarity index is considered a measure of goodness of fitting [13, 14]. When $Diss < 0.1$, a model fitting is considered good [13], or when $0.08 \leq Diss \leq 0.12$ it can be considered acceptable. Index $Diss$ indicates the proportion of respondents that should change their choice in order to get a perfect fitting [7]. While CUB model with shelter choice and CUBE model allow a *Diss* index, an index for CUB models with covariates cannot be derived. When comparing CUB models with covariates it should be compared likelihood measures between CUB model without and with covariates. The log-likelihood can be used to compare nested models [13] and all CUB model extensions are nested into the CUB model that estimates parameters ξ and π. In order to measure the goodness of a CUB model with covariates, we have to rely on log-likelihood comparison. Let us consider the relation

$$\ell(\theta_0) \leq \ell(\theta) \leq \ell(\theta_{sat}) \tag{1.11}$$

with $\ell(\theta_0)$ the likelihood of null model (only the constant), $\ell(\theta)$ the model estimated likelihood and $\ell(\theta_{sat})$ the saturated model likelihood [13]. This relation implies that greater is likelihood and better is an estimated model so that it make sense to compare a CUB model $(0,0)$, i.e. $\theta(\pi,\xi)$, with a CUB model (p,q), i.e. $\theta'' = (\beta_i, \gamma_j), i = 1, \ldots, p+1$ and $j = 1, \ldots, q+1$.

The Likelihood Ratio Test, $LRT = -2(\ell(\theta') - \ell(\theta''))$, is a measure of deviance between log-likelihoods of two models with one model nested in another. The probability distribution of LRT follows a χ^2 distribution with degrees of freedom equal to the difference between parameters of compared models, as it shows Table 1.1.

Table 1.1 CUB model comparisons

CUB models	Δ Log-likelihood	Degrees of freedom
$CUB(p,0)$ vs $CUB(0,0)$	$2(\ell_{10} - \ell_{00})$	p
$CUB(0,q)$ vs $CUB(0,0)$	$2(\ell_{01} - \ell_{00})$	q
$CUB(p,q)$ vs $CUB(0,0)$	$2(\ell_{11} - \ell_{00})$	$p+q$

Such a test gives an indication on how good is a nested model compared to the basic one. With respect to *shelter choice* and *overdispersion*, LRT should be taken into account the distribution χ^2 with 1 degree of freedom when comparing log-likelihoods. Moreover, because of a particular parametrization, the *p*-value should be halved [15, 17]. In the next sections will be described applications of CUB models to a case study on food packaging.

1.3 A Food Packaging Survey

In order to show an application of CUB models, we are going to describe a study on food packaging [2]. Consumers are usually overwhelmed by the great number of food packaging at grocery stores and packaging is considered a silent vendor as the first characteristic evaluated by customers [26, 27]. A questionnaire on food packaging was developed to have an overview on satisfaction and pitfalls when customers are facing with packaged food and buying foods at the grocery store. The study has been conducted in Italy and in Austria, collecting opinions and satisfaction grades of 209 subjects. The survey was centred on questions about packaging attributes like capacity to preserve food, resealability and easy peel. About the buying behaviour, subjects were asked to rate attention paid to several aspects, e.g. brand, packaging, price, etc. (Table 1.2) shows the main variables and coding of measurement scale.

Table 1.2 Variables and measurement scales

Variables	Δ Attributes	Coding
Attention paid to attributes	Nutrition facts, no GMO food, region of provenance, seasonality,	1 minimum attention
	brand, price, discounted price, innovation, advertisement, packaging	6 maximum attention
Satisfaction about packaging	Ability to preserve the food, resealability and easy peel	1 minimum satisfaction 10 maximum satisfaction
Opinions on packaging	Preservatives are the main responsible of freshness, packaging is the main responsible of freshness	1 no at all 10 definitely yes

Subjects were also asked demographic and habit information, some of them introduced as covariates (see Table 1.3) to improve the CUB model fitting. The survey included questions on attention level about some aspects related to products respondents usually buy at grocery stores (attention variables). Respondents were also asked to evaluate their satisfaction level about the following specific packaging characteristics: the ability to preserve foods, the resealability and easy peel. Moreover, they were asked to express opinions about packaging and preservatives that are usually involved in food preservation (satisfaction and opinion variables).

CUB models have been applied to attention variables and results are shown in Table 1.4. Results show parameter estimate, dissimilarity index and log-likelihood of each variable. Dissimilarity indexes are lower than 0.12 except for Advertisement. CUB model has two advantages: the first one is a model with only two parameters indicating we have a parsimonious model, the second one is a useful visual description by representing parameters into two dimensional space (Fig. 1.2).

Figure 1.2 indicates that the variable with the highest attention and the lowest *uncertainty* is *price*, while in the opposite position there is *innovation*. Whereas price and discounted price get the highest attention and the lowest *uncertainty*, respondents don't pay much attention to nutrition facts and innovation revealing very different behavior (high uncertainty). About *packaging*, they pay attention to it but *uncertainty* is high indicating very different attentional levels among respondents. Let us take a look at Table 1.5. CUB models with and without covariates are applied and Log-likelihoods are compared by Chi-square tests.

Table 1.3 Covariates for CUB model

Covariates	Δ Description	Coding
Sex	36% males	0 male
	64% females	1 female
Nationality	68% Italy	0 Italy
	32% Austria	1 Austria
Educational level	9.3% elementary	1 elementary
	28.1% intermediate	2 intermediate
	43.4% high school	3 high school
	19.2% graduate	4 graduate
Age	Min: 20	Continuous variable
	Max: 82	
Income (monthly in Euros)	26.3% < 800	$1 < 800$
	54.1% $800 - 1700$	$2800 - 1700$
	13.9% $1800 - 2900$	$31,800 - 2900$
	5.7% > 2900	$4 > 2900$
Purchase frequency	54% Rarely	0 rarely
	46% frequently	1 frequently
Attention paid to:		
Biodegradable packaging	66% yes	0 yes
	34% no	1 no
Resealable packaging	66.4% yes	0 yes
	33.6% no	1 no
Easy peel	55.6% yes	0 yes
	44.4% no	1 no

Table 1.4 CUB model estimates for attention variables

Coding	Variable name	$\pi(s.e.)$	$\xi(s.e.)$	Diss	$\ell_{(0,0)}$
1	Nutritional values	0.157(0.045)	0.990(0.054)	0.078	-321.810
2	No GMO food	0.254(0.094)	0.820(0.071)	0.109	-326.081
3	Provenance	0.387(0.069)	0.082(0.030)	0.057	-306.455
4	Provenance	0.345(0.073)	0.075(0.037)	0.093	-310.359
5	Brand	0.327(0.101)	0.485(0.054)	0.045	-326.487
6	Price	0.779(0.056)	0.146(0.017)	0.113	-265.980
7	Discounted price	0.610(0.061)	0.060(0.019)	0.106	-260.895
8	Innovation	0.053(0.066)	0.990(0.231)	0.087	-330.258
9	Advertisement	0.489(0.099)	0.864(.044)	0.166	-305.048
10	Packaging	0.177(0.098)	0.339(0.090)	0.064	-329.843

p-Values in Table 1.5 show that log-likelihoods for CUB models with covariates increase. Significant covariates can help to understand how behaviours of subject's subgroups differ in terms of attention paid to attributes. The introduction of covariates in the CUB model is supposed to be significant when log-likelihoods between CUB model without and with covariates are significantly different. Table 1.6 describes which direction takes $1 - \xi$ (attention) once covariates are introduced.

Fig. 1.2 Attention variables with increasing attention (feeling) and uncertainty when parameters tend to 1

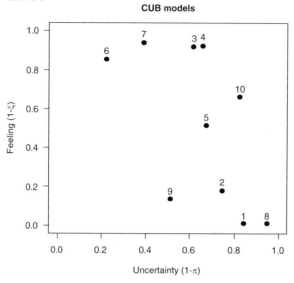

Table 1.5 Significant covariates for CUB models

Coding	Variable name	π	ξ	$2(\ell_{(p,q)} - \ell_{(0,0)})$	Df, p-value
1	Nutrition facts	Gender	Education	14.074	2, <0.0001
2	No GMO food	–	Nationality	4.174	1, <0.05
3	Provenance	–	Age	27.560	1, <0.0001
4	Seasonality	–	Age	29.586	1, <0.0001
5	Brand	–	Gender	4.582	1, <0.05
6	Price	–	Income	34.522	1, <0.0001
7	Discounted price	–	Income	32.574	1, <0.0001
8	Innovation	–	Age	19.560	1, <0.0001

Table 1.6 Attention direction for significant covariates

Variable name	Covariate	Coding	Attention
No GMO food	Nationality	0;1	Increase
Provenance	Age	20–82	Increase
Seasonality	Age	20–82	Increase
Brand	Gender	0;1	Decrease
Price	Income	1;4	Decrease
Discounted price	Income	1;4	Decrease
Innovation	Age	20–82	Decrease

In Table 1.6 we realize that older subjects paid more attention to provenance than the younger ones and males (coded as 0) paid more attention to brand than females. Another interesting results is that Italians seem to pay less attention to *no GMO food* than Austrian. Let us take a look at subgroups that come from crossing the categorical variables gender and education. We saw that covariate nutrition facts improve log-likelihoods of CUB models for all subgroups but with different implications. The logistic functions (1.4)–(1.5) estimate parameters π and ξ for each subgroup so that to have rating distribution estimations (Table 1.7).

The expected value $E(Y)$ is derived as follows (see [14]):

$$E(Y) = \pi(m-1)\left(\frac{1}{2} - \xi\right) + \frac{(m+1)}{2}. \tag{1.12}$$

Females have higher *uncertainty* than males and their responses present higher dispersion. About education, the higher education affect attention paid to nutrition facts. The expected values of females are quite similar whereas there is a clear

Table 1.7 Gender and education covariates for nutrition facts

Gender-education	$1 - \pi$	$1 - \xi$	$E(Y)$
Male-elementary	0.604	0	2.51
Female-elementary	0.844	0	3.11
Male-intermediate	0.604	0.0007	2.51
Female-intermediate	0.844	0.0007	3.11
Male-high school	0.604	0.047	2.61
Female-high school	0.844	0.047	3.15
Male-graduate	0.604	0.788	4.07
Female-graduate	0.844	0.788	3.73

gap between high school ($E(Y) = 2.61$) and graduate ($E(Y) = 4.07$) conditions for males. Figure 1.3 shows estimated probability distributions about covariates gender and education.

The distributions in left panel (females) are flatter than ones in right panel (males) indicating high *uncertainty*. Moreover, we cannot discriminate between elementary and intermediate because of overlapping distributions. Respondents with a degree clearly are connected with a higher level of attention to the nutrition facts. CUB models can take into account several choice behaviors and in order to give an example let us consider the variable advertisement. The dissimilarity index is quite high ($Diss = 0.166$) meaning a bad match between estimated and observed distributions. About observed relative frequencies of advertisement, we see a Mode at $y = 1$. The score 1 has a high frequency and could indicate a shelter choice at $c = 1$. The hypothesis is that some respondents have chosen the lowest score in order to simplify the task and in order to test it we apply the CUB model with shelter choice at $c = 1$ (Table 1.8).

Fig. 1.3 Estimated probability distributions of responses to nutrition facts for females (left panel) and males (right panel)

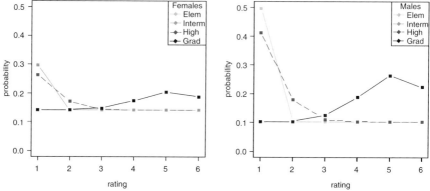

Table 1.8 Model without and with shelter for advertisement

Model	π(s.e.)	ξ(s.e.)	Diss. index	Log-likelihood
CUB	$\pi = 0.489(0.099)$	$0.864(0.044)$	0.166	-305.048
CUB+shelter	$\pi_1 = 0.504(0.075)$ $\pi_2 = 0.209(0.082))$ $\pi^* = 0.707(0.109)$	$0.621(0.037)$	0.021	-293.032

The parameter $\delta = 0.285(0.044)$ is significant and dissimilarity index decreases from 0.166 to 0.021. The CUB model with shelter choice improves the model and a shelter choice behavior seems to explain for an over-selection of score 1. Parameters ξ and π take different values in the CUB model with shelter choice so that we finally have higher attention (from 0.135 to 0.378) and lower *uncertainty* (from 0.511 to 0.292). About satisfaction and opinion variables, as introduced before, respondents evaluated their satisfaction concerning some packaging characteristics (ability to preserve foods, resealability and easy peel). Moreover they expressed opinions on packaging and preservatives revealing their beliefs. CUB models have been applied to satisfaction and opinion variables (Table 1.9) and results show acceptable fitting indexes for variables easy peel and preservatives.

Table 1.9 CUB model estimates for satisfaction and opinion variables

Coding	Variable name	π(s.e.)	ξ(s.e.)	Diss. index	$\ell_{(0,0)}$
1	Preservation	$0.695(0.065)$	$0.366(0.018)$	0.123	-384.280
2	Resealability	$0.668(0.066)$	$0.333(0.019)$	0.122	-383.851
3	Easy peel	$0.628(0.072)$	$0.418(0.021)$	0.085	-393.884
4	Preservatives	$0.536(0.071)$	$0.293(0.023)$	0.104	-399.914
5	Packaging	$0.553(0.074)$	$0.322(0.025)$	0.164	-400.412

Parameter estimates, as coordinates, describe points in Fig. 1.4 and suggests two clusters, the satisfaction variables (1, 2, 3) and the opinion variables (4, 5) with the second ones in the upper right corner of the space. Opinion variable positions suggest high uncertainty and high feeling.

Fig. 1.4 Satisfaction and opinion variables as coded in Table 1.9 with increasing feeling and uncertainty when parameters tend to 1

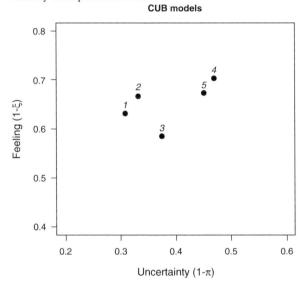

Dissimilarity indexes displayed in Table 1.9 indicate estimated distributions not fitted so well to the observed data. A possibility is to investigate for groups of subjects that behave in different ways by introducing covariates into CUB models. Results of CUB model with covariates are shown in Table 1.10.

Table 1.10 Significant covariates for CUB models

Coding	Variable name	π covariate	ξ covariate	$2(\ell_{(p,q)} - \ell_{(0,0)})$	Df, p-value
1	Preservation	Age	Purchase frequency	35.584	2, <0.0001
2	Resealability	–	Nationality Purchase frequency Resealable packaging	62.471	3, <0.0001
3	Easy peel	–	Nationality Attention to easy peel Resealable packaging	63.335	3, <0.0001
4	Preservatives	Resealability attention	Income	50.129	2, <0.0001
5	Packaging	–	Nationality Purchase frequency	37.641	2, <0.0001

The introductions of covariates led to an increment of log-likelihood and an improvement of CUB models. This result suggests that there are clusters with different grades of satisfaction or different opinions on how preservatives and packaging are linked to food preservation. Let us consider, for instance, satisfaction about food preservation that seems to be not homogeneous among respondents. Covariates *age* and *purchase frequency* were significant for *uncertainty* and *feeling* (satisfaction) respectively so that they outline subgroups with different grades of satisfaction. In order to estimate probability distributions of subgroups that come from crossing variables age and purchase frequency, we made age a discrete variable. Probability distributions are shown in Fig. 1.5.

Fig. 1.5 Estimated probability distributions about food preservation for frequent buyers of packaged products (left panel) and not frequent buyers (right panel)

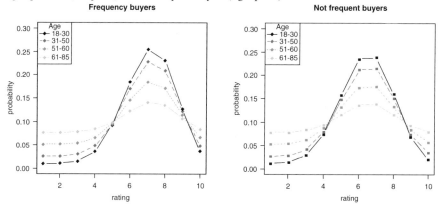

From Fig. 1.5 we see that age affects *uncertainty*: older the respondents flatter the distribution. Older respondents display greater uncertainty whereas frequent buyers seem to be more satisfied than not frequent buyers about packaged fresh foods.

1.4 Final Remarks

The main aim of CUB models was to explain the psychological mechanism underlying choice processes [10, 12]. Moreover, in order to take into account several choice behavior, model extensions have been developed [16]. Within the framework of preference evaluation methods, CUB models are considered a stated preference method and suited to many real cases [25, 7, 6, 20], confirming CUB models as useful and theorem based [19] statistical models. Moreover CUB models have been applied in combination and integration with other methods like conjoint analysis [3], indicating that they are also very flexible models. *feeling* and *uncertainty* as latent variables are supposed to be involved in the choice process of an item. The interpretation is wide with feeling indicating constructs (satisfaction, preference or attention) that are linked to the measurement scale adopted. In the real case study just described, CUB models have been applied to specific questions relating to the level of attention respondents usually pay to specific characteristics at grocery stores. Moreover questions regarded also satisfaction level and opinions about some food packaging characteristics. Results showed high levels of attention for price, discounted price, seasonality and provenance but with different levels of *uncertainty*. Packaging as a variable received medium-high grades of attention but *uncertainty* was high: respondents seem to pay very different levels of attention about packaging when they buy foods at supermarket. Packaging was not affected by any covariates so the high *uncertainty* could indicate an attribute (packaging) respondents are not used to evaluate or to consider when they buy products at the supermarket. This could mean a low knowledge of the real utility/importance of packaging. Some demographic characteristics introduced as covariates revealed the presence of clusters of subjects whose responses were quite different in terms of *feeling* and *uncertainty*. For example, males and females don't pay the same attention to brands or again the product provenance seems to be affected by age, older respondents are more interested product provenance than younger ones. The CUB model extension with shelter choice has been applied to variable advertisement. The model fitting improved a lot with a shelter at c = 1. From this results we reasonably state that respondents tend to simplify the answer. Maybe choosing a grade reflecting how much attention is paid to advertised products is not a simple task. Respondents were more satisfied with preservation of the food and the resealability of the packaging than with packaging with easy peel. Finally, from introducing covariates, we saw that a high frequency of packaged-food product purchasing and satisfaction for food preservation are linked in some ways. Concluding, the CUB models have proven to be very useful, flexible and constantly evolving. They have a wide range of possible applications not only as a single method but also in combination or integration with others statistical methods.

References

1. Alriksson, S., Öberg, T.: Conjoint analysis for environmental evaluation. A review of methods and applications. Environ. Sci. Pollut. Res. **15**(3), 244–257 (2008)
2. Arboretti Giancristofaro, R., Bordignon, P.: Combination of Uniform Binomial (CUB) Models: an application to the evaluation of food packaging. In: JSM Proceedings, Statistical Computing Section, pp. 2581–2593. American Statistical Association, Alexandria (2014)
3. Arboretti, R., Bordignon, P.: Consumer preferences in food packaging: CUB models and conjoint analysis. Br. Food J. **118**(3), 527–540 (2016)

4. Box, G.E.P., Draper, N.R.: Empirical Model Building and Response Surfaces. Wiley, New York (1987)
5. Capecchi, S., Endrizzi, I., Gasperi, F., Piccolo, D.: A multi-product approach for detecting subjects and objects' covariates' in consumer preferences. Br. Food J. **118**(3), 515–526 (2016)
6. Cicia, G., Corduas M., Del Giudice T., Piccolo, D.: Valuing consumer preferences with the CUB model: a case study of fair trade coffee. Int. J. Food Syst. Dyn. **1**, 82–93 (2010)
7. Corduas, M., Iannario, M., Piccolo, D.: A class of statistical models for evaluating services and performances. In: Monari, P., Bini, M., Piccolo, D., Salmaso, L. (eds.), Statistical Methods for the Evaluation of Educational Services and Quality of Products, pp. 99–117. Physica-Verlag HD, Heidelberg (2009)
8. Corduas, M., Cinquanta, L., Ievoli, C.: The importance of wine attributes for purchase decisions: a study of Italian consumers' perception. Food Qual. Prefer. **28**, 407–418 (2013)
9. D'Elia, A.: Il meccanismo dei confronti appaiati nella modellistica per graduatorie: sviluppi statistici ed aspetti critici. Quad. Stat. **2**, 173–203 (2000)
10. D'Elia, A.: A mixture model with covariates for ranks data: some inferential developments. Quad. Stat. **5**, 1–25 (2003)
11. D'Elia, A.: New developments in ranks data modelling with covariate. Atti della XLII Riunione Scientifica SIS, pp. 233–244. CLEUP, Padova (2004)
12. D'Elia, A., Piccolo, D.: A mixture model for preferences data analysis. Comput. Stat. Data Anal. **49**, 917–934 (2005)
13. Iannario, M.: Fitting measures for ordinal data models. Quad. Stat. **11**, 39–72 (2009)
14. Iannario, M.: On the identifiability of a mixture model for ordinal data. METRON Int. J. Stat. **LXVIII**(1), 87–94 (2010)
15. Iannario, M. Modelling shelter choices in a class of mixture models for ordinal responses. Stat. Method Appl. **21**, 1–22 (2012)
16. Iannario, M.: A class of model for ordinal data analysis: statistical issues and new developments. Mixture and latent variable models for casual inference and analysis of socio-economic data, FIRB meeting, Perugia (2013)
17. Iannario, M.: Modelling uncertainty and overdispersion in ordinal data. Commun. Stat. Theory Method **43**, 771–786 (2014)
18. Iannario, M., Piccolo, D.: CUB models: statistical methods and empirical evidence. In: Kennet, S.R., Salini, S. (eds.) Modern Analysis of Customer Satisfaction Surveys: With Applications Using R, pp. 231–258. Wiley, Chichester (2012)
19. Iannario, M., Piccolo, D.: A theorem on CUB models for rank data. Stat. Probab. Lett. **91**, 27–31 (2014)
20. Iannario, M., Manisera, M., Piccolo, D., Zuccolotto, P.: Sensory analysis in the food industry as a tool for marketing decisions. Adv. Data Anal. Classif. **6**, 303–321 (2012)

21. Kennet, S.R., Salini, S.: Modern analysis of customer satisfaction surveys: comparison of models and integrated analysis. Appl. Stoch. Model. Bus. Ind. **27**, 465–475 (2011)
22. Louviere, J.J., Hensher, D.A., Swait, J.D.: Stated Choice Methods. Analysis and Applications. Cambridge University Press, Cambridge (2000)
23. Piccolo, D.: On the moments of a mixture of uniform and shifted binomial random variables. Quad. Stat. **5**, 85–104 (2003)
24. Piccolo, D.: Computational issues in the E-M algorithm for ranks model estimation with covariates. Quad. Stat. **5**, 1–22 (2003)
25. Piccolo, D., D' Elia A.: A new approach for modelling consumers preferences. Food Qual. Prefer. **19**, 247–259 (2008)
26. Rundh, B.: Packaging design: creating competitive advantage with product packaging. Br. Food J. **111**(9), 988–1002 (2009)
27. Silayoi, P., Speece, M.: The importance of packaging attributes: a conjoint analysis approach. Eur. J. Mark. **41**, 1495–1517 (2007)

Chapter 2
Customer Satisfaction Heterogeneity

The measurement of the customer satisfaction concerns the gap between the customer expectations about the product or service and the perceptions of the customer after the consumption or use. In other words, the customer satisfaction is closely related to the concept of "perceived quality". According to the definition of Montgomery [24], it depends on how much the products or services meet the requirements of the consumers/users and it is directly connected to the homogeneity of the performance of the production process or service provision process.

When the performance is represented by a numerical variable, for instance when the customers express their satisfaction degree with a numerical score, quality is inversely related to variability. High variability of the process means inhomogeneous outputs and great percentage of waste. As a matter of fact, the process capacity indices are measures used to evaluate the ability of the process to produce little waste and are an inverse function of the standard deviation. Other important tools for process quality control are the so called control charts and among them the R charts and the S charts are commonly used to control range and standard deviation as main indices of variability.

In many cases the customer satisfaction is measured through categorical judgments, by using a Likert scale or a set of ordered categorical evaluations. Often, the number of possible levels used to represent the different satisfaction degrees are 4, 5, 7 or 10. Even if sometimes, in the customer satisfaction questionnaires, the possible response alternatives are linked to integer numbers which represent the ranks of the evaluations with respect to the judgment scale, the nature of the information provided by the answers of the respondents is categorical. For this reason, range, variance, standard deviation and other indices used to measure the variability of the satisfaction, after transformation of the ordinal assessments into numeric scores, often are not the suitable way to measure satisfaction inhomogeneity. The transformation of the original data can change the information provided by the statistical surveys and cause a bias in the estimation of the customer satisfaction parameters,

R. Arboretti et al., *Parametric and Nonparametric Statistics for Sample Surveys and Customer Satisfaction Data*, SpringerBriefs in Statistics, https://doi.org/10.1007/978-3-319-91740-5_2

leading to unreliable results because the transformed data do not reflect the real customers opinions.

Hence, the application of suitable statistical techniques for categorical variables is preferable to the transformation of data into numeric scores and consequent application of inferential methods for variability parameters (e.g. F test for variance comparison). In this framework, interesting methodological tools to measure satisfaction inhomogeneity are represented by the indices of heterogeneity for categorical variables, such as Gini's index, Shannon's entropy, Rényi's family of measures and others. These indices can be used for inferential purposes, concerning heterogeneity estimation and test of hypothesis for comparing the heterogeneity of two or more populations. Another good reason why the use of indices of heterogeneity for categorical variables is preferable to data transformation and application of indices of variability, is that the detection of customer groups characterized by "within homogeneity" and "between heterogeneity" could be a useful starting point of market segmentation and product/service differentiation strategies.

The present chapter is dedicated to the description of testing methods, for comparing the (categorical) satisfaction heterogeneities of two or more customer populations. These methods are based on a nonparametric approach and on the comparison of the Pareto diagrams of the probability distributions. In the next section the heterogeneity of categorical variables is defined and suitable indices of heterogeneity are presented. In the section after that, the theory of the two-sample test for heterogeneity comparisons is introduced and some real applications shown. The final section includes the extension of the method under study to the comparison of more than two populations, from both theoretical and application point of view.

2.1 Heterogeneity Indices

The notion of statistical heterogeneity for categorical data finds several applications in different disciplines such as genetics, physics, engineering, environmental sciences, sociology, economics, and others. According to the problem and the scientific framework, it is associated to the concept of diversity, entropy, mutability, dispersion, differentiation, dissimilarity or uniform distribution. For instance, it can be used for studying market segmentation [10], biodiversity [26, 32, 33, 7], process capability [11], clustering [38], genetic differentiation [6], customer satisfaction [5] and many other phenomena typical of the mentioned disciplines.

Let us assume that the random variable X_j, representing the customer satisfaction of population j ($j = 1, \ldots, C; C \geq 2$), may take m categories (judgments) $v_1, v_2, \ldots, v_m$ and f_{jh} denotes the absolute frequency of the j-th sample for the h-th category ($j = 1, \ldots, C; h = 1, \ldots, m$). A $2 \times m$ contingency table of the absolute frequencies $[f_{jh}]$ is observed. In other words the categorical response variable X_j takes categories in $\{v_1, \ldots, v_m\}$, with unobserved probability distribution $\Pr\{X_j = v_h\} = \pi_{jh}$, $h = 1, \ldots, m$.

The heterogeneity of X_j or, equivalently, the heterogeneity of its distribution, is minimum (in other words the homogeneity is maximum) when in the j-th population one modality is observed with probability 1 (certain event) and all the other modalities with probability 0 (impossible event). In this population there is full judgment homogeneity and then the distribution is degenerate. Conversely the heterogeneity of X_j is maximum when all the modalities/categories are observed with the same probability. In this population the distribution is uniform over the set of modalities, that is $\pi_{jh} = 1/m, \forall h$. Thus, heterogeneity depends on the concentration of probabilities over the categories $v_1, v_2, \ldots, v_m$.

An suitable measure of heterogeneity $\eta_j = het(X_j)$ must satisfy the following properties:

1. it takes its minimum when the distribution is degenerate, i.e. when there is an integer $r \in \{1, \ldots, m\}$ such that $\pi_{jr} = 1$ and $\pi_{jh} = 0, \forall h \neq r$;
2. the farther the distribution from the degenerate case and the closer to the uniform case, the greater the index value;
3. it takes its maximum in case of uniform distribution, i.e. $\pi_{jh} = 1/m, \forall h \in \{1, \ldots, m\}$.

The properties listed above hold for many different index types (for a review see [31]). Each of this indices can be used as a measure of the degree of heterogeneity. By assuming that $\log(\cdot)$ correspond to the natural logarithm and that $0 \cdot \log(0) = 0$, let us consider the following indices:

Shannon entropy ([35]): $\eta_j^{(S)} = -\sum_{h=1}^{m} \pi_{jh} \log(\pi_{jh})$,

Gini heterogeneity ([17]): $\eta_j^{(G)} = \sum_{h=1}^{m} \pi_{jh}(1 - \pi_{jh}) = 1 - \sum_{h=1}^{m} \pi_{jh}^2$,

Leti index ([20]): $\eta_j^{(L)} = \prod_{h=1}^{m} (\pi_{jh})^{-\pi_{jh}}$,

Frosini index - euclidean distance ([16]): $\eta_j^{(Fe)} = \sqrt{\sum_{h=1}^{m}(\pi_{jh} - 1/m)^2}$,

Frosini index - Manhattan distance ([16]): $\eta_j^{(Fm)} = \sum_{h=1}^{m} |\pi_{jh} - 1/m|$,

Rényi index of order δ ([34]): $\eta_j^{(R_\delta)} = \frac{1}{1-\delta} \log \sum_{h=1}^{m} \pi_{jh}^\delta$.

To facilitate the interpretation of the indices, after suitable transformation, it is possible to compute, for each of them, the normalized version, which takes the minimum value 0 in case of degenerate distribution and the maximum value 1 in case of uniform distribution (see [4]).

The index of entropy proposed by Shannon, is equal to 0 in the former case and equal to $\log(m)$ in the latter case. Hence the normalized version of this index is equal to

$$\tilde{\eta}_j^{(S)} = \frac{\eta_j^{(S)}}{\log(m)}.$$

Also the Gini's index, in case of degenerate distribution, takes value 0. When the distribution is uniform it is equal to $(m-1)/m$. Thus the normalized version is

$$\tilde{\eta}_j^{(G)} = \frac{m}{m-1} \cdot \eta_j^{(G)}.$$

The Gini's index can be interpreted as arithmetic mean of $1 - \pi_{jh}$ $(h = 1, \cdots, m)$, which can be considered measures of heterogeneity of the single attributes v_h $(h = 1, \cdots, m)$. The logic underlying the index of Leti is similar, because the Leti's measure of heterogeneity corresponds to the geometric mean of $1/\pi_{jh}$ $(h = 1, \cdots, m)$. This index is a monotonic transformation of Shannon's entropy since $\eta_j^{(L)} = \exp(\eta_j^{(S)})$. Its normalized version is

$$\tilde{\eta}_j^{(L)} = \frac{\eta_j^{(L)} - 1}{m - 1}.$$

Quite a general approach was followed by Frosini, whose indices of homogeneity are defined as distances between the vector of probabilities $(\pi_{j1}, \pi_{j2}, \ldots, \pi_{jm})'$, which characterize the distribution of X_j (observed relative frequencies in descriptive statistics), and the vector of expected probabilitites in case of uniform distribution $(1/m, 1/m, \ldots, 1/m)'$. The normalized version of the index computed as euclidean distance is

$$\tilde{\eta}_j^{(Fe)} = 1 - \sqrt{\frac{m}{m-1} \cdot \eta_j^{(G)2}},$$

which is an increasing function of the Gini's index. It can be proved that $\tilde{\eta}_j^{(Fe)} = 1 - \sqrt{1 - \tilde{\eta}_j^{(G)}}$. Both indices are a decreasing function of $\sum_h \pi_{jh}^2$, which is commonly used as measure of homogeneity by several contributions.

The normalized Frosini's index based on Manhattan distance is

$$\tilde{\eta}_j^{(Fm)} = 1 - \frac{m}{2(m-1)} \cdot \eta_j^{(Fm)}.$$

The generalized index of entropy defined by Rény is a non-increasing function of parameter δ, with $\delta \neq 1$. Some of the indices presented above, or increasing transformations of them, can be considered members of the Rény's family of indices. For instance, let us consider the following cases:

- $\eta_j^{(R_1)} = \lim_{\delta \to 1} \left[\frac{1}{1-\delta} \log(\sum_{h=1}^m \pi_{jh}^\delta) \right] = \eta_j^{(S)}$,

- $\eta_j^{(R_2)} = -\log \sum_{h=1}^m \pi_{jh}^2 = -\log(1 - \eta_j^{(G)})$,

- $\eta_j^{(R_\infty)} = \lim_{\delta \to \infty} \left[\frac{1}{1-\delta} \log(\sum_{h=1}^m \pi_{jh}^\delta) \right] = -\log[\sup_h(\pi_{jh})]$.

Table 2.1 shows the values of some of the most common indices of heterogeneity (in both non-normalized and normalized versions) related to specific probability distributions in the case of 4 modalities or satisfaction levels. Three aspects are evident from the table: the monotonic relationship between index value and degree of heterogeneity, the similar normalized values of $\eta^{(G)}$ and $\eta^{(S)}$ and the normalized values of $\eta^{(R_\infty)}$, which are much different from those of the other indices.

2.2 Two-Sample Test for Dominance in Heterogeneity

2.2.1 Problem Definition

Let us consider the two-sample test where the hypothesis under study consists in the comparison between the heterogeneities of the customer satisfactions of two populations, by assuming that the satisfaction is represented by the categorical variable X_j $(j = 1, 2)$, with support given by the set of modalities $\{v_1, v_2, \ldots, v_m\}$. We could be interested in assessing the plausibility of the hypothesis that the heterogeneity of one population is greater than that of the other (one-sided test) or the hypothesis that the two heterogeneities are not equal (two-sided test). To this end, let us suppose that one random sample from each of the two populations is selected and denote the size of sample j with N_j $(j = 1, 2)$. As indicated in the previous section, the probability distribution of X_j is $\{\pi_{jh}, h = 1, \ldots, m\}$, with $\pi_{jh} = Pr\{X_j = v_h\} \geq 0$ and therefore $\sum_h \pi_{jh} = 1$. The probabilities π_{jh} are unknown parameters of the two populations.

If we denote the heterogeneity degree of the judgments of the j-th population with $het(X_j)$, the hypotheses of the one-sided testing problem can be formally defined as follows:

$$H_0 : het(X_1) = het(X_2)$$

against

$$H_1 : het(X_1) > het(X_2).$$

Table 2.1 Probability distributions in the case of $m = 4$ modalities and measures of heterogeneity

Modalities				Index				Normalized index			
v_1	v_2	v_3	v_4	$\eta^{(S)}$	$\eta^{(G)}$	$\eta^{(R_3)}$	$\eta^{(R_\infty)}$	$\tilde{\eta}^{(S)}$	$\tilde{\eta}^{(G)}$	$\tilde{\eta}^{(R_3)}$	$\tilde{\eta}^{(R_\infty)}$
1.00	0.00	0.00	0.00	0.00	0.00	0.00	0.00	0.00	0.00	0.00	0.00
0.70	0.20	0.10	0.00	0.46	0.80	0.52	0.36	0.61	0.58	0.38	0.26
0.50	0.30	0.15	0.05	0.64	1.14	0.93	0.69	0.85	0.82	0.67	0.50
0.40	0.30	0.20	0.10	0.70	1.28	1.15	0.92	0.93	0.92	0.83	0.66
0.30	0.25	0.25	0.20	0.75	1.38	1.36	1.20	0.99	0.99	0.98	0.87
0.25	0.25	0.25	0.25	0.75	1.39	1.39	1.39	1.00	1.00	1.00	1.00

Let us denote with $\pi_{j(1)} \geq \pi_{j(2)} \geq \cdots \geq \pi_{j(m)}$ the ordered probabilities of the j-th population. All the indices defined in the previous section are *order invariant*, i.e. their values do not change if they are computed with the ordered probabilities instead of the unordered ones. Formally, if we denote with η_j any index to measure $het(X_j)$ and we indicate it as a function of the unknown parameters, we can state that

$$\eta_j(\pi_{j(1)}, \pi_{j(2)}, \ldots, \pi_{j(m)}) = \eta_j(\pi_{j1}, \pi_{j2}, \ldots, \pi_{jm}),$$

thus we can express $het(X_j)$ by using the ordered probabilities.

Let us observe that:

$$\pi_{1(h)} = \pi_{2(h)} \forall h = 1, 2, \ldots, m \Rightarrow het(X_1) = het(X_2).$$

In other words, two populations with the same ordered distributions, are equally heterogeneous. Moreover, if $\pi_{1(h)} = \pi_{2(h)} \forall h = 1, 2, \ldots, m$, exchangeability holds and the permutation testing principle applies. According to it, the null hypothesis of the testing problem can be represented as:

$$H_0 : \pi_{1(h)} = \pi_{2(h)} \forall h = 1, 2, \ldots, m$$

or equivalently as

$$H_0 : \Pi_{1(h)} = \Pi_{2(h)} \forall h = 1, 2, \ldots, m - 1,$$

where $\Pi_{j(h)} = \sum_{s=1}^{h} \pi_{j(s)}$ are the cumulative ordered probabilities, with $\Pi_{j(m)} = \sum_{h=1}^{m} \pi_{j(h)} = 1$.

The alternative hypothesis of the problem can be written as:

$$H_1 : \Pi_{1(h)} \leq \Pi_{2(h)} \forall h = 1, 2, \ldots, m - 1,$$

and the strict inequality holds for at least one $h = 1, 2, \ldots, m - 1$.

The problem is then similar to stochastic dominance for ordinal categorical variables, where the order is here determined according to the values of the parameters π_{jh}, and not according to the categories $v_1, v_2, \ldots, v_m$. In other words, under the null hypothesis, the Pareto diagrams of the probabilities distributions are coincident and under the alternative one diagram dominates the other. Therefore the problem under study can also be named *test on dominance in heterogeneity* and be defined in terms of comparison of probability concentrations.

2.2.2 Permutation Test

For problems of stochastic dominance many exact and approximate solutions have been proposed in the literature (see [1, 18, 19, 22, 21, 23, 25]). For the univariate case several authors proposed solutions based on the restricted maximum likelihood ratio test [15, 36, 40]. According to these proposals, under the null and alternative hypothesis, the distributions of the test statistics asymptotically are mixtures of chi-squared whose weights essentially depend on the unknown population distribution. Nonparametric solutions are proposed by other authors [39, 13, 27, 28, 29].

For comparing the heterogeneities of two populations, it is reasonable to choose, as test statistic, the difference between the sampling values of an index η, as follows

$$T_\eta = \hat{\eta}_1 - \hat{\eta}_2,$$

where $\hat{\eta}_j = \eta_j(p_{j(1)}, p_{j(2)}, \ldots, p_{j(m)})$, $p_{j(h)} = f_{j(h)}/N_j$, with $f_{j(h)}$ and $p_{j(h)}$ ordered relative frequencies and ordered absolute frequencies respectively, observed on the two samples ($h = 1, \ldots, m; j = 1, 2$). The ordered relative frequencies $p_{j(h)} = \hat{\pi}_{j(h)}$ are point estimates of the unknown ordered probabilities $\pi_{j(h)}$ as well as the sample indices $\hat{\eta}_j$ represent suitable point estimates of the indices η_j. The null hypothesis is rejected in favour of the alternative for large values of the test statistic.

In order to compute the p-value $\lambda = Pr\{T_\eta \geq T_\eta^{(obs)}|H_0\}$, where $T_\eta^{(obs)}$ is the observed value of the test statistic, we need to know the sampling null distribution of the test statistic. Arboretti et al. [6] proposed a permutation solution that assumes exchangeability under the null hypothesis. The authors consider different options as test statistic, considering alternatively the indices of Gini, Shannon and Rény of order 3 and ∞. The Rény's index of order 2 is the most commonly used but, according to the property mentioned above, it is permutationally equivalent to the index of Gini.

Exchangeability under H_0 holds if the true order of the unknown probabilities were known. Since in practice the true order is not known, it must be estimated with the sample data. Hence exchangeability under H_0 is not exact, because the order of the probabilities is estimated with the order of the observed frequencies (empirical order) which presents a random component and it is subjected to the sampling variations. Thus the solution is data driven. Anyhow, estimated and true order are asymptotically equal with probability one, according to the Glivenko-Cantelli theorem [37], thus exchangeability holds asymptotically and the permutation test is approximate for finite sample sizes.

For each sample, the observed relative frequencies are sorted in decreasing order, as well as in the Pareto diagram, and the table of ordered relative frequencies $p_{j(h)}$ (ordered table) is determined. The original variables X_j are then transformed in such a way that, within the j-th sample, v_h is replaced by the integer number r if $p_{jh} = p_{j(r)}$. The observed value of the test statistic $T_\eta^{(obs)}$ is computed as a function of the frequencies of the ordered table. The null distribution of the test statistic is obtained by considering all the permutations of the transformed dataset or (for computational convenience) a random sample of them. In the latter case the permutation p-value is not exact but estimated with the conditional Monte Carlo method. For each permutation a new (permuted) table of the ordered frequencies is obtained and the corresponding permutation value of the test statistic is computed. The permutation p-value is then

$$\lambda_\eta = \frac{\sharp(T_\eta^{*(b)} \geq T_\eta^{(obs)}|X)}{B},$$

where $T_\eta^{*(b)}$ is the permutation value of the test statistic corresponding to the b-th permutation, B is the number of permutations and the numerator is the number of permutation values of the test statistic greater than or equal to the observed one, given the observed dataset X. Thus the inferential result depends on the space generated by the permutations of X, that is the orbit associated with X. As usual, the null hypothesis is rejected when $\lambda_\eta \leq \alpha$, where α is the significance level of the test.

Let us consider the following practical example, where 350 customers of a food outlet were interviewed to evaluate a sweet snack bought in that shop. The evaluation consisted in choosing a categorical judgment within a Likert scale with seven satisfaction levels, where level 1 corresponds to "not at all satisfied" and level 7 corresponds to "very much satisfied".

Table 2.2 Contingency table of observed absolute frequencies in the customer satisfaction survey about a sweet snack

Satisfaction	Gender		Total
	Male	Female	
1 = not at all	18	5	23
2	25	17	42
3	31	28	59
4	35	36	71
5	27	39	66
6	22	28	50
7 = very much	17	22	39
Total	175	175	350

We wish to test the hypothesis that the satisfaction of males is more heterogeneous than that of females at significance level $\alpha = 0.05$. The observed contingency table is shown in Table 2.2.

It is evident that the sample modal satisfaction for males corresponds to level 4 and for females to level 5. Hence, from a descriptive point of view and by using the mode as location index, the satisfaction of females for the sweet snack seems to be greater than that of males.

Since we are interested in the comparison of the heterogeneities, we can consider the Pareto diagrams of the frequencies distributions, where the two cumulative ordered frequency polygons can be represented to compare the frequency concentrations (see Fig. 2.1).

The frequencies of the customer satisfaction distribution for males seem to be less concentrated, supporting the hypothesis of greater heterogeneity of judgements for this category. The computation of the sample indices of heterogeneity of Gini,

Fig. 2.1 Pareto diagrams of the customer satisfaction survey about a sweet snack

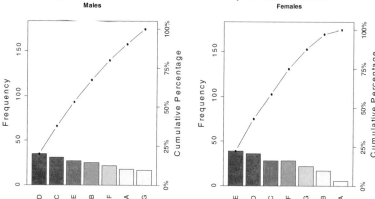

Table 2.3 Sample indices of heterogeneity in the customer satisfaction survey about a sweet snack

Index	Gender	
	Male	Female
Gini	0.849	0.831
Shannon	1.916	1.835
Rényi-3	1.862	1.739
Rényi-∞	1.609	1.501

Shannon, Rény of order 3 and Rény of order ∞, conforms this idea as shown in Table 2.3.

To test whether the observed positive differences between the indices of the two groups are significant, we apply the permutation test described above, by estimating the permutation p-values with $B = 10,000$ conditional Monte Carlo simulations and using all the four measures of heterogeneity just mentioned. Table 2.4 reports the observed values of the test statistics and the corresponding p-values:

According to the table, the observed difference of the sample indices is (positively) significant when we use the indices of Gini, Shannon and Rény of order 3. It is non-significant when we use the index of Rény of order ∞. Thus, according to the first three procedures, the heterogeneity of males' judgements is greater than the heterogeneity of females. This result is consistent with some evidences of the literature.

Table 2.4 One-sided test on heterogeneity in the customer satisfaction survey about a sweet snack

Index	$T_\eta^{(obs)}$	p-value
Gini	0.018	0.028
Shannon	0.081	0.012
Rényi-3	0.123	0.046
Rényi-∞	0.108	0.226

Arboretti et al. [6] prove that, in general, the test based on the index of Rényi-∞ is the least powerful among the four considered, and this is more evident for high degrees of heterogeneity. As a matter of fact the degrees of heterogeneity in the problem presented are very high. The normalized values of the Gini's indices for males and females are 0.991 and 0.970 respectively; the normalized values of the Shannon's indices are 0.985 and 0.943; for the index of Rényi-3 we have 0.957 and 0.894 and for the index of Rényi-∞ the values are 0.827 and 0.771.

2.3 Two-Sided Test

Sometimes, in the two-sample problem, the interest of the study concerns the test of a two-sided hypothesis. For example let us consider the case of a survey to evaluate the satisfaction of students about professors' teaching effectiveness in an academic course. The students were divided into two groups that attended the lectures separately at different times. A a matter of fact, the course was repeated twice and the same classes were held by the same professors separately for the two groups.

Table 2.5 Contingency table of observed absolute frequencies of students' satisfaction about teaching effectiveness

Satisfaction	Group		Total
	A	B	
Very dissatisfied	5	12	17
Moderately dissatisfied	18	35	53
Moderately satisfied	28	26	54
Very satisfied	9	9	18
Total	60	82	142

As shown in Table 2.5, the satisfaction is expressed by using a 4-level scale. The satisfaction levels are: *Very dissatisfied, Moderately dissatisfied, Moderately satisfied* and *Very satisfied*. The table shows the observed absolute frequencies of the judgements for group A, group B and for the pooled sample. We are interested in comparing the two distributions, to test whether the heterogeneities of the group satisfactions are different at the significant level $\alpha = 0.01$.

Fig. 2.2 Pareto diagrams of students' satisfaction about teaching effectiveness

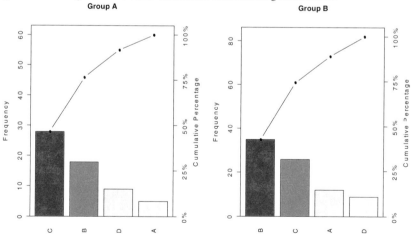

The Pareto diagrams of the data presented in Table 2.5 are shown in Fig. 2.2. From a descriptive point of view, to determine which of the two heterogeneities is greater is not simple. The sample values of the heterogeneity indices can help in this assessment.

For the two-sided test, Arboretti and Bonnini [2] and Arboretti et al. [4] propose the application of the permutation solution described in the previous section by using, as a test statistic, the difference squared of the sample indices. Formally the problem can be represented by the null hypothesis

$$H_0 : het(X_1) = het(X_2)$$

against the alternative hypothesis

$$H_1 : het(X_1) \neq het(X_2),$$

where X_1 and X_2 represent the satisfaction of the two compared groups. By denoting with $\tilde{\eta}_j$ a given sample index of heterogeneity, the test statistic for this problem is

$$T_\eta = (\tilde{\eta}_1 - \tilde{\eta}_2)^2,$$

and large values of T_η lead to the rejection of the null hypothesis in favour of the alternative. Under the null hypothesis, exchangeability holds, even if, as remarked above, it is approximate for finite sample sizes and exact only asymptotically. Thus the permutation distribution of T_η can be estimated, the p-value of the test computed and the usual decision rule applied to decide whether reject or not the null hypothesis. In the mentioned publications, a further index, consistent with the two-sided nature of the alternative hypothesis, is taken into account as alternative to the other indices listed above. This is the well known chi-squared index

$$\eta_j^{(\chi^2)} = \sum_{h=1}^m \frac{(Np_{jh} - \frac{N}{m})^2}{\frac{N}{m}}.$$

However, the test based on the chi-squared index is not well approximated as well as the test statistics based on Shannon's and Gini's indices, because it tends to be anticonservative under the null hypothesis. Furthermore it is less powerful than the other two tests under the alternative hypothesis.

Table 2.6 Sampe indices of heterogeneity of students' satisfaction about teaching effectiveness

Index	Group	
	A	B
Gini	0.663	0.684
Shannon	1.209	1.251
Rényi-3	1.010	1.085
Rényi-∞	0.762	0.851

According to Table 2.6, the heterogeneity of judgments by group B seems to be lightly higher, because all the sample indices of this group are lightly greater. We apply the permutation method here described to test the significance of the differences between the sample indices.

Table 2.7 Two-sided test on heterogeneity of students' satisfaction about teaching effectiveness

Index	$T_\eta^{(obs)}$	p-value
Gini	0.021	0.624
Shannon	0.043	0.627
Rényi-3	0.075	0.615
Rényi-∞	0.089	0.672

In Table 2.7 we can see that all the p-values are much greater than the significance level, hence we cannot reject the null hypothesis of equal satisfaction heterogeneities.

2.4 Multisample Test

When the compared categorical variables are more than two, a sort of *ANOVA* can be applied to test the hypothesis that the heterogeneities of the $C \geq 3$ distributions are not equal (see [3]). An example is provided in [14], where the results of a customer satisfaction survey on facilitis services in Terminal 2 at Tampere Airport are published.

Table 2.8 Contingency table of observed absolute frequencies of customer satisfaction about facilities services in Terminal 2 at Tampere Airport

Satisfaction	Terminal area		
	Entrance concourse	Departure lounge	Arrival lounge
Highly dissatisfied	4	4	3
Somewhat dissatisfied	7	9	9
Neutral	42	35	48
Somewhat satisfied	50	50	38
Highly satisfied	27	32	32

The data of the frequency distributions of the customer satisfaction about facilities services are reported in Table 2.8. The total number of customers, travellers who took part to the customer satisfaction survey, is 130 for all the areas under evaluation, i.e. "Entrance course", "Departure lounge" and "Arrival lounge". We are interested in comparing the heterogeneities of the judgements for the three areas ($\alpha = 0.10$).

In generale, given C distributions with $C \geq 3$, the C-sample testing problem with null hypothesis of equality in heterogeneity of the C distributions and alternative hypothesis of inequality in heterogeneity of some of them, can be formally written as

$$H_0 : het(X_1) = het(X_2) = \cdots = het(X_C)$$

versus

$$H_1 : \exists (i, j) | het(X_i) \neq het(X_j), i \neq j = 1, 2, \cdots, C.$$

The alternative hypothesis states that at least one couple of categorical random variables which represents customer satisfactions present different heterogeneities. Even in the multisample case, the hypotheses of the problem can be defined by comparing the cumulative ordered probabilities, that is the Pareto diagrams of the probability distributions, as follows:

$$H_0 : \Pi_{1(h)} = \Pi_{2(h)} = \cdots = \Pi_{C(h)}, \forall h = 1, 2, \cdots, m-1$$

versus

$$H_1 : \exists (i,j) | \Pi_{i(h)} \neq \Pi_{j(h)} \text{ for some } h = 1, 2, \cdots, m-1, i \neq j = 1, 2, \cdots, C,$$

with $\Pi_{j(h)} = \sum_{s=1}^{h} \pi_{j(s)}$, where $\pi_{j(s)}$ is the s-th ordered probability in the j-th distribution, that is in the probability distribution of the categorical variable X_j.

If the null hypothesis is true, asymptotically exact or, for finite sample, approximate exchangeability holds. Thus the permutation distribution of a suitable test statistic can be estimated and the p-value can be computed or estimated as usual. A suitable test statistic for the multisample problem is:

$$T_\eta = \sum_{j=1}^{C} (\hat{\eta}_j - \hat{\eta}_\bullet)^2,$$

where $\hat{\eta}_j$ is a sample index of heterogeneity for the j-th sample and $\hat{\eta}_\bullet$ is the same index of heterogeneity computed for the pooled sample. The algorythm for the execution of the test is the same described above, except for the test statistic and the number of samples.

Fig. 2.3 Pareto diagrams of customer satisfaction about facilities services in Terminal 2 at Tampere Airport

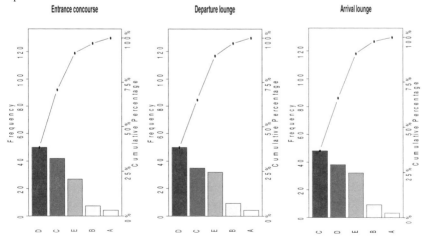

In the presented application example, from the descriptive point of view, the comparison of the sample Pareto diagrams does not reveal evident differences in the frequency concentration over the categories (judgements) among the three samples (see Fig. 2.3).

Table 2.9 Sampe indices of heterogeneity of customer satisfaction about facilities services in Terminal 2 at Tampere Airport

Index	Terminal area		
	Entrance concourse	Departure lounge	Arrival lounge
Gini	0.700	0.713	0.712
Shannon	1.323	1.358	1.344
Rényi-3	1.152	1.195	1.201
Rényi-∞	0.956	0.956	0.996

The values of the sample indices of heterogeneity, as shown in Table 2.9, seem to be consistent with the hypothesis of equality in heterogeneity. The described permutation test can be applied to test whether the differences of the sample indices between samples are significant.

Table 2.10 Multisample test on heterogeneity of customer satisfaction about facilities services in Terminal 2 at Tampere Airport

Index	$T_\eta^{(obs)}$	p-value
Gini	0.025	0.786
Shannon	0.043	0.820
Rényi-3	0.105	0.737
Rényi-∞	0.320	0.366

Since the p-values of all the tests, reported in Table 2.10, are greater than α, than we cannot reject the null hypothesis, that is there is not empirical evidence in favour of the hypothesis of inequality in heterogeneity between some of the three compared distributions.

2.5 Further Theoretical Developments and Practical Suggestions for Users

The permutation test described in this chapter can be applied by choosing one of the several indices of heterogeneity. In general the index of Shannon and the index of Gini are preferable because the corresponding tests are well approximated (rarely anticonservative under the null hypothesis) and in many cases more powerful than other tests. These tests are unbiased, consistent and the convergence to one of the power is fast.

Bonnini [8] proposed a nonparametric alternative method for testing heterogeneity comparisons based on a different resampling strategy. This proposal consists in a bootstrap test, where the null distribution of the test statistic is computed or estimated through resamplings with replacements of the observed sample data, instead of resamplings without replacements typical of the permutation tests. According to the simulation study presented in this paper, the power behavior of the bootstrap test is similar (in general slightly less powerful) to that of the permutation test but sometimes it is anticonservative under H_0, in particular for small sample sizes and low heterogeneities, when the test statistic is based on $\eta^{(R_\infty)}$. In general it doesn't happen when the Shannon's index or the Gini's index are used, thus these two measures of heterogeneity are preferable even with the bootstrap test. The bootstrap test can be used instead of the permutation test in particular for very small sample sizes, because it allows a larger cardinality of the permutation space. Furthermore the bootstrap techniques are appropiate when the inferential goal of the study concerns interval estimation of the differences between the heterogeneities of two or more populations. The slight power loss that occurs with the bootstrap test respect to the permutation one, usually does not occur when using the index $\eta^{(R_3)}$.

When there is not an evident reason to prefer one of the several indices of heterogeneity, a combined test can be applied by considering the problem as a multiaspect test (see [30]). As a matter of fact, the several measures of heterogeneity are not equivalent and different indices can lead to different results. Thus, each of them can be considered as a partial aspect of a complex phenomenon and a combined test seems to be appropriate. Following this idea, Arboretti et al. [7] proposed to use, for the two sample problem, the following test statistic, based on the sum of the sample normalized indices of Gini, Shannon, Rényi-3 and Rényi-∞

$$T_{Dcomb} = T_{\eta^{(G)}} + T_{\eta^{(S)}} + T_{\eta^{(R_3)}} + T_{\eta^{(R_\infty)}},$$

where $T_{\eta^{(G)}} = \tilde{\eta}_1^{(G)} - \tilde{\eta}_2^{(G)}$, $T_{\eta^{(S)}} = \tilde{\eta}_1^{(S)} - \tilde{\eta}_2^{(S)}$ and $T_{\eta^{(R_\delta)}} = \tilde{\eta}_1^{(R_\delta)} - \tilde{\eta}_2^{(R_\delta)}$.
This can be considered a direct additive combination of the four tests. Other combining rules based on the combination of p-values are applied by Bonnini (14b) [9]. These methods, in particular by using Fisher's and Liptak's combination, are a good compromise that often garantees rejection rates not greater than α under H_0 and a good power under H_1, by compensating the p-values of the partial tests, even when some of these are anticonservative or underpowered.

A description of the permutation methods for heterogeneity comparisons and of suitable R codes for their practical implementation, are presented in Bonnini et al. [12].

References

1. Agresti, A., Klingenberg, B.: Multivariate tests comparing binomial probabilities, with application to safety studies for drugs. Appl. Stat. **54**, 691–706 (2005)

2. Arboretti Giancristofaro, R., Bonnini, S.: Permutation tests for heterogeneity comparisons in presence of categorical variables with application to university evaluation. Adv. Methodol. Stat. **4**, 1, 21–36 (2007)
3. Arboretti Giancristofaro, R., Bonnini, S., Pesarin, F.: Comparisons of heterogeneity: a nonparametric test for the multisample case. In: Lopez-Fidalgo, J., Rodriguez-Diaz, J.M., Torsney, B. (eds.) mODa 8 – Advances in Model-Oriented Design and Analysis, pp. 17–24. Physica-Verlag, Heidelberg (2007)
4. Arboretti Giancristofaro, R., Bonnini, S., Pesarin, F., Salmaso, L.: One-sided and two-sided nonparametric tests for heterogeneity comparisons. Statistica **LXVIII**(1), 57–69 (2008)
5. Arboretti Giancristofaro, R., Bonnini, S., Salmaso, L.: Employment status and education/employment relationship of PhD graduates from the University of Ferrara. J. Appl. Stat. **36**(12), 1329–1344 (2009)
6. Arboretti, G.R., Bonnini, S., Pesarin, F.: A permutation approach for testing heterogeneity in two-sample problems. Stat. Comput. **19**, 209–216 (2009)
7. Arboretti, G.R., Bonnini, S., Corain, L., Vidotto, D.: Environmental odor perception: testing regional differences on heterogeneity with application to odor perceptions in the area of Este (Italy). Environmetrics **26**(6), 418–430 (2015)
8. Bonnini, S.: Testing for heterogeneity for categorical data: permutation solution vs. bootstrap method. Commun. Stat. A Theor. **43**(4), 906–917 (2014)
9. Bonnini, S.: Combined tests for comparing mutabilities of two populations. In: Topics in Statistical Simulation. Book of Proceedings of the Seventh International Workshop on Simulation 2013, Rimini, 21–25 May 2013, pp. 67–78. Springer, New York (2014)
10. Bonnini, S.: Multivariate approach for comparative evaluations of customer satisfaction with application to transport services. Commun. Stat. Simul. C **45**(5) (2016). https://doi.org/10.1080/03610918.2014.941685
11. Bonnini, S.: Nonparametric test on process capability. In: Cao, R., Gonzalez Manteiga, W., Romo, J. (eds.) Nonparametric Statistics, Proceedings of the second ISNPS Conference, Cadiz, June 2014, pp. 11–18. Springer, Cham (2016)
12. Bonnini, S., Corain, L., Marozzi, M., Salmaso, L.: Nonparametric Hypothesis Testing. Rank and Permutation Methods with Applications in R. Wiley, Chichester (2014)
13. Brunner, E., Munzel, U.: The nonparametric Behrens-Fisher problem: asymptotic theory and small-sample approximation. Biom. J. **42**, 17–25 (2000)
14. Chumakova, A.: Customer satisfaction on facility services in terminal 2 of Tampere Airport. Bachelor's thesis. Tampere University of Applied Sciences, Degree Programme in Tourism, Tampere (2014)
15. Cohen, A., Kemperman, J.H.B., Madigan, D., Sarkrowitz, H.B.: Effective directed tests for models with ordered categorical data. Aust. N. Z. J. Stat. **45**, 285–300 (2000)
16. Frosini, B.V.: Heterogeneity indices and distances between distributions. Metron, **XXXIX**, 3–4 (1981)
17. Gini, C.: Variability and mutability. In: Legal-Economic Studies of the Faculty of Law, University of Cagliari (1912)

18. Han, K.E., Catalano, P.J., Senchaudhuri, P., Mehta, C.: Exact analysis of dose-response for multiple correlated binary outcomes. Biometrics **60**, 216–224 (2004)
19. Hirotsu, C.: Cumulative chi-squared statistic as a tool for testing goodness-of-fit. Biometrika **73**, 165–173 (1986)
20. Leti, G.: Entropy, a Gini index and other heterogeneity measures. Metron **XXIV**, 1–4 (1965)
21. Loughin, T.M.: A systematic comparison of methods for combining p-values from independent tests. Comput. Stat. Data Anal. **47**,467–485 (2004)
22. Loughin, T.M., Scherer, P.N.: Testing for association in contingency tables with multiple column responses. Biometrics **54**, 630–637 (1998)
23. Lumley, T.: Generalized estimating equations for ordinal data: a note on working correlation structures. Biometrics **52**, 354–361 (1996)
24. Montgomery, D.C.: Introduction to statistical quality control. Wiley, Hoboken (2013)
25. Nettleton, D., Banerjee, T.: Testing the equality of distributions of random vectors with categorical components. Comput. Stat. Data Anal. **37**, 195–208 (2001)
26. Patil, G.P., Taillie, C.: Diversity as a concept and its measurement (with discussion). J. Am. Stat. Assoc. **77**, 548–567 (1982)
27. Pesarin, F.: Goodness-of-fit testing for ordered discrete distributions by resampling techniques. Metron **LII**, 57–71 (1994)
28. Pesarin, F.: Multivariate Permutation Test With Application to Biostatistics. Wiley, Chichester (2001)
29. Pesarin, F., Salmaso, L.: Permutation tests for univariate and multivariate ordered categorical data. Aust. J. Stat. **35**, 315–324 (2006)
30. Pesarin, F., Salmaso, L.: Permutation Tests for Complex Data: Theory, Applications and Software. Wiley, Chichester (2010)
31. Piccolo, D.: Statistica. Il Mulino, Bologna (2000)
32. Pielou, E.C.: Ecological Diversity. Wiley, New York (1975)
33. Pielou, E.C.: Mathematical Ecology. Wiley, New York (1977)
34. Rényi, A.: Calculus des probabilités. Dunod, Paris (1966)
35. Shannon, C.E.: A mathematical theory of communication. Bell Syst. Tech. J. **27**, 379–423, 623–656 (1948)
36. Silvapulle, M.J., Sen, P.K.: Constrained Statistical Inference, Inequality, Order, and Shape Restrictions. Wiley, New York (2005)
37. Shorack, G.R., Wellner, J.A.: Empirical Processes with Applications to Statistics. Wiley, New York (1986)
38. Thornton-Wells, T.A., Moore, J.,H., Haines, J.L.: Dissecting trait heterogeneity: a comparison of three clustering methods applied to genotypic data. BMC Bioinf. **7**, 204 (2006)
39. Troendle, J.F.: A likelihood ratio test for the nonparametric Behrens-Fisher problem. Biom. J. **44**(7), 813–824 (2002)
40. Wang, Y.: A likelihood ratio test against stochastic ordering in several populations. J. Am. Stat. Assoc. **91**, 1676–1683 (1996)

Chapter 3
Ranking Multivariate Populations

The need to define an appropriate ranking of several populations of interest, i.e. processes, products, and so on is very common within many areas of applied research such as Food Science, Chemistry, Engineering, Biomedicine, etc. The idea of ranking in fact occurs more or less explicitly any time when in a study the goal is to determine an ordering among several input conditions/treatments with respect to one or more outputs of interest when there might be a "natural ordering". We remark that the "natural ordering" should be referred to the way in which the response is interpreted and not to any kind of a priori knowledge on ordering of populations that is not assumed at all. This happens very often in the context of food sensory analysis problems where the populations can be varieties, products, processes, etc. and the inputs are for example the food-related physicochemical properties which are put in relation with some suitable outputs such as any performance measure. At the same time, the ranking problem is a typical interdisciplinary subject, just think for example on the development process of a new product where technological issues and statistical techniques are jointly involved in order to achieve high quality and potentially successful products. Many times the populations of interest are multivariate in nature, meaning that many aspects of that populations can be simultaneously observed on the same unit/subject. For example, in many food-related experiments the treatments under evaluation provide an output of several univariate responses. From a statistical point of view, when the response variable of interest is multivariate in nature, the inferential problem may become quite difficult to cope with, due to the large dimensionality of the parametric space. Moreover, when the goal is that of comparing several multivariate populations, a further element of difficulty is related to the nature of the response variable. If we consider a continuous response, provided that the underlying distributional and sampling assumptions are met and the degree of freedom are large enough, then inference on populations can be performed using classical methods (e.g. such as Hotelling T2). But when the response variables

R. Arboretti et al., *Parametric and Nonparametric Statistics for Sample Surveys and Customer Satisfaction Data*, SpringerBriefs in Statistics, https://doi.org/10.1007/978-3-319-91740-5_3

are ordered categorical the difficulties of the traditional methods based on contingency tables may become insurmountable. Nonparametric inference based on the
NPC—NonParametric Combination of several dependent permutation test statistics
(see [10]), allows us to overcome most of these limitations, without the necessity
of referring to assume any specified random distribution. The main advantages of
using the permutation and combination approach to classify and rank several multivariate populations is that it is the only one testing method which allow us to derive
multivariate directional p-values that can be calculated also when the number of
response variables are much more larger than the number of replications (so-called
finite-sample consistency of combined permutation tests). It is worth noting that in
this situation, which can be common in many real applications, all traditional parametric and nonparametric testing procedures are not appropriate at all. For deeper
introduction on the topic of ranking of multivariate populations we refer the reader
to Corain and Salmaso [5].

3.1 Ranking Methods

Following results in Arboretti er al.[1], let us assume that data were drawn from each
of C multivariate populations $\Pi_1, \ldots, \Pi_C$ (i.e. items/groups/treatments), $C > 2$, by
means of a sampling procedure, so as to make inference on their possible equality
and in case of rejection of this hypothesis to classify those populations in order to
obtain a relative ranking from the 'best' to the 'worst' according to a pre-specified
meaningful criterion. We use the term relative ranking because we want to underline that it is not an absolute ranking but a kind of ordering that is only refereed
to the C populations at hand. Let Y be the p-dimensional response variable represented as a p-vector of the observed data from population Π and let us assume,
without loss of generality, that large values of each univariate aspect Y correspond to
a better marginal performance, so that when comparing two populations the possible marginal stochastic superiority should result in a high ranking position. In other
words, we are assuming the criterion "the larger the better". The term "large values" has a clear meaning in case of continuous responses, while in case of binary or
ordered categorical responses, this should be intended in term of "large proportion"
and of "large frequencies of high score categories" respectively. The marginal univariate components of Y are not restricted to belong to the same type, in other words
we can consider also the situation of mixed variables (some continuous/binary and
some others ordered categorical). We recall that our goal is to classify and ranking
$\Pi_1, \ldots, \Pi_C$ multivariate populations with respect to p marginal variables where are
available C samples, from each one population, of n_j independent replicates represented by the random variables $Y_1, .., Y_C$, $j = 1, \ldots, C$. In other words we are looking
for an estimate $\hat{r}(\Pi_j)$ of the rank $r(\Pi_j)$, i.e. the relative stochastic ordering of each
population when compared among all other populations, i.e. more formally

$$r_j = r(\Pi_j) = 1 + \sum_{j \neq h} I(Y_j \overset{d}{<} Y_h) = 1 + \{ \#Y_j \overset{d}{<} Y_h \}, j, h = 1, \ldots, C, j \neq h \quad (3.1)$$

where $I(\cdot)$ is the indicator function and # means the number of times. This definition extends into a nonparametric multivariate framework the traditional definition of ranking, hence it is consistent with the ranking problem literature (see [7] and [8]). Under the hypothesis of distributional equality of the C populations, all true global rankings would necessarily be equal to one, hence they would be in a full ex-aequo situation, that is

$$r(\Pi_j|H_0) = \{1 + \#Y_j \overset{d}{<} Y_h, h = 1,\ldots,C, j \neq h\} = 1, \forall j \qquad (3.2)$$

This situation of equal ranking where all populations belong to just one ranking class may be formally represented in a testing-like framework where the hypotheses of interest are:

$$\begin{cases} H_0 : Y_1 \overset{d}{=} Y_2 \overset{d}{=} \ldots \overset{d}{=} Y_C \\ H_1 : \exists Y_j \overset{d}{\neq} Y_h, j, h = 1,\ldots,C, j \neq h \end{cases} \qquad (3.3)$$

In case of rejection of the global multivariate hypothesis H_0, that is when data are evidence of the fact that at least one population behaves differently from the others, it is of interest to perform inferences on pairwise comparisons between populations, i.e.

$$\begin{cases} H_{0(jh)} : Y_j \overset{d}{=} Y_h \\ H_{1(jh)} : Y_j \overset{d}{\neq} Y_h, j, h = 1,\ldots,C, j \neq h \end{cases} \qquad (3.4)$$

Note that a rejection of at least one hypothesis $H_{0(jh)}$ implies that we are not in an equal ranking situation, that is at least one multivariate population has a greater ranking position than some others. Note that, as usual in the framework of the C-sample inference, the rejection of the global null hypothesis is not informative on the specific alternative has caused the rejection so that post-hoc analysis is needed to look for which alternative is more likely. In this connection, to make inference on which marginal variable(s) that inequality is mostly due to, it is useful considering the inferences on univariate pairwise comparisons between populations, defined as:

$$\begin{cases} H_{0k(jh)} : Y_{jk} \overset{d}{=} Y_{hk} \\ H_{1k(jh)} : (Y_{jk} \overset{d}{<} Y_{hk}) \cup (Y_{jk} \overset{d}{>} Y_{hk}), j, h = 1,\ldots,C, j \neq h \end{cases} \qquad (3.5)$$

because when $Y_{jk} \overset{d}{\neq} Y_{hk}$ is true, then one and only one between $Y_{jk} \overset{d}{<} Y_{hk}$ and $Y_{jk} \overset{d}{>} Y_{hk}$ is true, i.e. they cannot be jointly true. Looking at the univariate alternative hypothesis $H_{1k(jh)}$, note that we are mostly interested in deciding whether a population is either greater or smaller than another one (not only establishing if they are different). In this connection, we can take into account separately of the directional type alternatives, namely those that are suitable for testing both one-sided alternatives (see [10, p. 163] and [2]). Let $p_{k(jh)}^+$ and $p_{k(jh)}^-$ be the permutation-based marginal directional p-value statistics related to the stochastic inferiority or superiority alternatives $H^+_{1k(jh)} : Y_{jk} \overset{d}{>} Y_{hk}$ and $H^-_{1k(jh)} : Y_{jk} \overset{d}{<} Y_{hk}$, respectively. Since by definition $p^+_{k(jh)} = 1 - p^-_{k(jh)} = p^-_{k(hj)}$, note that all one-sided inferential results related to the hypotheses (3.5) can be represented as follows:

$$
P^+ = \begin{bmatrix}
- & p^+_{1(1,2)} & p^+_{1(1,3)} & \cdots & p^+_{1(1,C)} \\
p^+_{1(2,1)} & - & p^+_{1(2,3)} & \cdots & p^+_{1(2,C)} \\
\cdots & \cdots & - & \cdots & \cdots \\
p^+_{1(C-1,1)} & p^+_{1(C-1,2)} & \cdots & - & p^+_{1(C-1,C)} \\
p^+_{1(C,1)} & p^+_{1(C,2)} & \cdots & p^+_{1(C,C-1)} & -
\end{bmatrix}
$$

$$
= \begin{bmatrix}
- & p^+_{p(1,2)} & p^+_{p(1,3)} & \cdots & p^+_{p(1,C)} \\
p^+_{p(2,1)} & - & p^+_{p(2,3)} & \cdots & p^+_{p(2,C)} \\
\cdots & \cdots & - & \cdots & \cdots \\
p^+_{p(C-1,1)} & p^+_{p(C-1,2)} & \cdots & - & p^+_{p(C-1,C)} \\
p^+_{p(C,1)} & p^+_{p(C,2)} & \cdots & p^+_{p(C,C-1)} & -
\end{bmatrix}
$$

Finally, let be $p^+_{\bullet(j,h)}$ the directional p-value statistics calculated via nonparametric combination methodology [10]. All the $C \times (C-1)$ $p^+_{\bullet(j,h)}$ can be represented as follows:

$$
P^+_\bullet = \begin{bmatrix}
- & p^+_{\bullet(1,2)} & p^+_{\bullet(1,3)} & \cdots & p^+_{\bullet(1,C)} \\
p^+_{\bullet(2,1)} & - & p^+_{\bullet(2,3)} & \cdots & p^+_{\bullet(2,C)} \\
\cdots & \cdots & - & \cdots & \cdots \\
p^+_{\bullet(C-1,1)} & p^+_{\bullet(C-1,2)} & \cdots & - & p^+_{\bullet(C-1,C)} \\
p^+_{\bullet(C,1)} & p^+_{\bullet(C,2)} & \cdots & p^+_{\bullet(C,C-1)} & -
\end{bmatrix}
$$

Now, let α be the chosen significance α-level and let S the $C \times C$ matrix which transforms the adjusted (by multiplicity) p-values $p^+_{\bullet(j,h)adj}$ into 0-and-1 scores where each element $s_{j,h}$ takes the value of 0 if $p^+_{\bullet(j,h)adj} > \alpha/2$, otherwise it takes 1 if $p^+_{\bullet(j,h)adj} \leq \alpha/2$, that is

$$
S = \begin{bmatrix}
- & S_{(1,2)} & S_{(1,3)} & \cdots & S_{(1,C)} \\
S_{(2,1)} & - & S_{(2,3)} & \cdots & S_{(2,C)} \\
S_{(3,1)} & S_{(3,2)} & - & \cdots & S_{(3,C)} \\
\cdots & \cdots & \cdots & - & \cdots \\
S_{(C,1)} & S_{(C,2)} & \cdots & S_{(C,C-1)} & -
\end{bmatrix}
$$

In practice, S is nothing more than a synthetic representation of results from all multivariate directional pairwise comparisons suitable for estimating the possible pairwise dominances. If we consider either the sum of the $s(j,h)$ $0-1$ scores along the h-th column or the j-th row, then we are respectively counting the number of populations which, at the chosen significance α-level, are considered to be stochastically larger or smaller than the h-th population or the j-th row. That is, we are defining an estimate $\hat{r}(\Pi_h)$ and $\hat{r}(\Pi_j)$ of the rank $r(\Pi_h)$ or $r(\Pi_j)$, i.e. the relative stochastic ordering of each population when compared with all other populations, i.e. more formally

$$\hat{r}_h^D = 1 + \sum_{j=1}^{C} S_{(j,h)}, h = 1, \ldots, C \tag{3.6}$$

$$\hat{r}_j^U = 1 + \{\#(C - \sum_{h=1}^{C} S_{(j,h)}) > (C - \sum_{h=1}^{C} S_{(j',h)}), j', j = 1, \ldots, C; j \neq j' \tag{3.7}$$

where D and U and stands for downward and upward rank estimates respectively. We note that the ranking estimators defined in (3.6) and (3.7) are deriving by counting, on the basis of empirical evidence, how many populations are significantly stochastically larger/smaller than the h-th/j-th population at the chosen significance α-level. The two estimated rankings $\hat{r}^D$ and $\hat{r}^U$ of the true rank r are intentionally denoted with a different notation in order to highlight that sometimes they could provide different rank estimates for the same population. We remark that, as we outlined in our literature review, the use of a pairwise matrix as to derive a ranking is quite common, especially in the algorithmic ranking literature.

3.2 Set-Up of the Multivariate Ranking Problem

In Sect. 3.1 we formalized our approach to solve the problem we called the multivariate ranking problem, i.e. that of ranking several multivariate populations from the 'best' to the 'worst' according to a given pre-specified criterion when a sample from each population is available and for each marginal univariate response there is a natural preferable direction. Since the key element of our solution is a testing procedure suitable for multivariate one-sided alternatives, the NPC methodology represents our main methodological reference framework. In fact, to the best of our knowledge, the nonparametric combination of dependent permutation tests, the so-called NPC Tests, is the only method proposed by the literature suitable to achieve this goal. Moreover, when deriving the multivariate one-sided p-values we can also benefit from the flexibility of the method for obtaining a series of advantages: NPC methodology allows to handle with all type of response variable, i.e. numeric, binary and ordered categorical even in the presence of missing data (at random or not at random, i.e. non-informative or informative) and this can be done also when the number of response variables are much more larger than that of units without the need of having to worry about the curse of dimensionality or the problem of the reduction of degrees of freedom. On the contrary, thanks to the so-called finite-sample consistency of combined permutation tests, the power function does not decrease for any added variable which makes larger standardized noncentrality. It is worth noting that in this situation, which can be common in many real applications, all traditional parametric and nonparametric testing procedures are not at all appropriate (also in the case all multivariate alternatives were of two-sided type). Finally, the NPC approach has a lot of nice feature: it is very low demanding in terms of assumptions and provides always an exact solution for whatever finite sample size whenever the permutation principle applies, i.e. when the null hypothesis implies data exchangeability. We recall that our goal is to classify and ranking C multivariate populations with respect to several marginal variables where samples from each population are available. Note that the multivariate ranking problem is essentially related to a posthoc comparative multivariate C-sample problem where the populations of interest are treatments or groups or items to be investigated by an experimental or observation study.

It is worth noting that within the NPC framework an optimal statistic cannot exist because it is function of the population distributions which is unknown by definition [10]. For this reason it is important to consider for each type of response variable a number of different test statistics. We recall that each univariate partial test statistic we are presenting must be suitable for one-sided alternatives with respect to the hypotheses $H_{0k(jh)}$ vs. $H_{1k(jh)}$. When the univariate marginal response variable is continuous or binary, within the permutation framework we can use a number of test statistics suitable for one-sided alternatives. In this context, we underline that the test statistics should obviously be not permutationally equivalent, for example we can refer to the difference of sample means:

$$T_{DM,k(jh)} = \sum_i Y_{ijk}/n_j - \sum_i Y_{ihk}/n_h. \tag{3.8}$$

When the univariate marginal response variable is ordered categorical with S ordinal categories, within the permutation framework we can use a number of test statistics suitable for directional alternatives. Some examples of suitable test statistics are the Anderson-Darling test statistics:

$$T_{AD,k(jh)} = \sum_{s=1}^{S-1} N_{hsk} \cdot [N_{\cdot sk} \cdot (n_{jk} + n_{hk} - N_{\cdot sk})]^{-\frac{1}{2}} \tag{3.9}$$

where $N_{\cdot sk} = N_{jsk} + N_{hsk}$ are the cumulative frequencies, and the Multi-Focus test statistics:

$$T_{MF,ks(jh)} = (f_{jks} - \hat{f}_{jks})^2, s = 1, \ldots, S \tag{3.10}$$

where f_{jks} and $\hat{f}_{jks}$ are respectively the observed and the estimated frequencies of the s-th two-by-to sub-table; note that there is a number of S multi-focus statistics for each univariate response variable so that an additional combination phase is needed to obtain the k-th test statistic. In order to define one-sided multivariate test statistics within the combination of dependent permutation testing methodology, a suitable combining function must be chosen [10]. Frequently used combining functions are:

- Fisher combination: $\phi_F = -2\sum_k \log(\lambda_k)$
- Tippet combination: $\phi_T = \max_{1 \leq k \leq p}(1 - \lambda_k)$
- Direct combination: $\phi_D = \sum_k T_k$
- Liptak combination: $\phi_L = \sum_k \Phi^{-1}(1 - \lambda_k)$

where $k = 1, \ldots, p$ and Φ is the standard normal c.d.f.

3.3 Application to Food Sensory Analysis

Sensory analysis is a quantitative statistical subject aimed at using human senses (sight, smell, taste, touch and hearing) for the purposes of evaluating consumer products. The discipline requires a group of panellist, i.e. panels of human assessors, on whom the products are evaluated. By applying suitable statistical techniques to the results of a sensory test it is possible to make inferences and find out insights about the products under test. Especially in the food industry, useful experimental performance indicators are individual sensorial evaluations provided by trained people (panellists) during a so-called sensory test [9].

3.3.1 Wine Quality

The following sensory study aims at investigating the relation of wine quality sensory assessments with the related wine physicochemical properties on a large sample of red variants of the Portuguese "Vinho Verde" wine. This dataset is public available for research, the details are described in Cortez et al. [6]. The dataset is made by 1599 records referring to independent red wine testing sessions where at least three evaluations were made by wine experts. Each expert graded the wine quality between 0 (very bad) and 10 (very excellent) and, as final output of wine quality, the related median of scores was considered. Moreover, for each one tested red wine, a number of 11 physicochemical properties were recorded:

1. fixed acidity (tartaric acid—g/dm^3)
2. volatile acidity (acetic acid—g/dm^3)
3. citric acid (g/dm^3)
4. residual sugar (g/dm^3)
5. chlorides (sodium chloride—g/dm^3)
6. free sulfur dioxide (mg/dm^3)
7. total sulfur dioxide (mg/dm^3)
8. density (g/cm^3)
9. pH
10. sulphates (potassium sulphate—g/dm^3)
11. alcohol (% by volume)

First of all we represented in Fig. 3.1 the scatterplots of quality scores (median of scores provided by experts) versus each physicochemical property, along with the fitted line calculated by the simple linear regression.

According to the sign of the fitted regression lines, we classified the physicochemical properties into two main domains, i.e. the direct and the inverse set of quality-related physicochemical properties. In the fist domain there are "fixed acidity", "citric acid", "residual sugar", "sulphates" and "alcohol". In the second domain we found "volatile acidity", "chlorides", "free sulfur dioxide", "total sulfur dioxide",

Fig. 3.1 Scatterplot of quality vs. physicochemical property

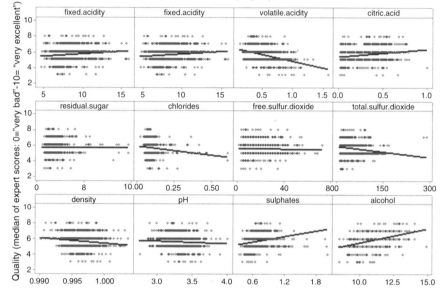

"density" and "pH". After that, we defined four quality groups according to the values of the quality score:

- Group Quality=1: scores equal or less to 4; 63 records met this condition;
- Group Quality=2: scores equal to 5; 681 records met this condition;
- Group Quality=1: scores equal to 6; 638 records met this condition;
- Group Quality=1: scores equal or greater to 7; 217 records met this condition.

Figure 3.2 displays the box plots of some of the most correlated properties by group property.

As expected, when we move from group 1 to 4 the physicochemical property increases or decreases accordingly to the positive or negative sign of the fitted regression line. We set up the ranking analysis problem as follows:

- type of design: one-way MANOVA design (four independent samples, i.e. four multivariate populations to be ranked);
- domain analysis: yes (two domains, direct and inverse quality-related physicochemical properties);
- type of response variables: numerical continuous (11 numerical responses);
- Ranking rule: "the higher the better" and "the lower the better" for the first and second domain respectively;
- Combining function: Fisher;
- B—Number of permutations: 2000;
- Significance α-level: 0.01.

Fig. 3.2 Boxplot of "volatile acidity", "density", "sulphates" and "alcohol" vs. Group Quality

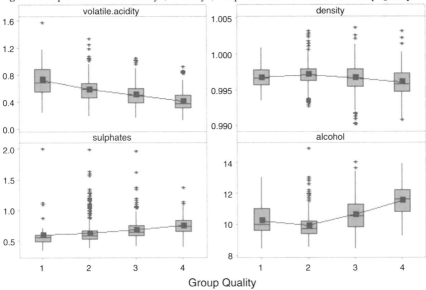

When applying the ranking method to wine quality data using directional parametric p-values performed via $B = 2000$ permutations we obtain results reported in Tables 3.1 and 3.2, where directional multivariate permutation p-values in Table 3.1 have been adjusted by using the Bonferroni-Holm-Shaffer method [11].

It is worth noting that ranking analysis results allow to strongly support that the sensorial perceived quality can be associated with higher or lower values of specific direct and inverse quality-related physicochemical properties.

3.3.2 Cream Cheese

The goal of this sensory study it to investigate the influence of fat content on sensory properties and consumer perception of dairy products of ten cream cheese. From the full set of ten products investigated in Bro et al. [4] we selected a subset set of five products whose description is detailed in Table 3.3.

From the full list of sensory descriptors as in Bro et al. [4], we focused on a subset set of six sensory descriptors, belonging to two main domain i.e. firmness and creaminess (Table 3.4).

Sensory evaluations was performed in three replicates by a panel consisting of eight panellists. Figures 3.3 and 3.4 show the boxplot and the sample mean of sensory data by each sensory descriptor and product.

Table 3.1 Direct quality-related physicochemical properties: global ranking results ($\alpha = 1\%$) and related adjusted directional multivariate and directional (raw-unadjusted) univariate permutation p-values for the Wine Quality sensory study ($\alpha = 1\%$)

Direct quality-related physicochemical properties			
Global ranking	4	3	2	1
Group Quality	1	2	3	4
1	–	0.636	1.000	1.000
2	0.003	–	1.000	1.000
3	0.001	0.001	–	1.000
4	0.001	0.001	0.000	–
Fixed acidity			
Group Quality	1	2	3	4
1	–	0.931	0.981	1.000
2	0.70	–	0.976	1.000
3	0.020	0.024	–	1.000
4	0.000	0.000	0.001	–
Citric acid			
Group Quality	1	2	3	4
1	–	1.000	1.000	1.000
2	0.000	–	0.997	1.000
3	0.000	0.004	–	1.000
4	0.000	0.000	0.000	–
Residual sugar			
Group Quality	1	2	3	4
1	–	0.179	0.142	0.531
2	0.821	–	0.268	0.949
3	0.859	0.733	–	0.966
4	0.471	0.052	0.035	–
Sulphates			
Group Quality	1	2	3	4
1	–	0.906	1.000	1.000
2	0.097	–	1.000	1.000
3	0.000	0.000	–	1.000
4	0.000	0.000	0.001	–
Alcohol			
Group Quality	1	2	3	4
1	–	0.002	0.999	1.000
2	0.999	–	1.000	1.000
3	0.002	0.000	–	1.000
4	0.000	0.000	0.000	–

Table 3.2 Inverse quality-related physicochemical properties: Global ranking results ($\alpha = 1\%$) and related adjusted directional multivariate and directional (raw-unadjusted) univariate permutation p-values for the Wine Quality sensory study ($\alpha = 1\%$)

Inverse quality-related physicochemical properties			
Global ranking 3	3	2	1
Group Quality 1	2	3	4
1 –	0.003	0.001	0.001
2 0.003	–	0.001	0.001
3 0.177	1.000	–	0.000
4 1.000	1.000	1.000	–

Volatile acidity			
Group Quality 1	2	3	4
1 –	0.000	0.000	0.000
2 1.000	–	0.000	0.000
3 1.000	1.000	–	0.000
4 1.000	1.000	1.000	–

Chlorides			
Group Quality 1	2	3	4
1 –	0.315	0.052	0.002
2 0.686	–	0.003	0.000
3 0.950	0.998	–	0.000
4 0.998	1.000	1.000	–

Free sulfur dioxide			
Group Quality 1	2	3	4
1 -	1.000	0.999	0.917
2 0.000	–	0.012	0.000
3 0.002	0.988	–	0.020
4 0.088	1.000	0.980	–

Total sulfur dioxide			
Group Quality 1	2	3	4
1 -	1.000	0.980	0.512
2 0.000	–	0.000	0.000
3 0.021	1.000	–	0.005
4 0.490	1.000	0.996	–

Density			
Group Quality 1	2	3	4
1 –	0.981	0.386	0.013
2 0.020	–	0.000	0.000
3 0.615	1.000	–	0.001
4 0.988	1.000	1.000	–

(continued)

Table 3.2 (continued)

Inverse quality-related physicochemical properties				
pH				
Group Quality	1	2	3	4
1	–	0.000	0.001	0.000
2	1.000	–	0.942	0.086
3	1.000	0.059	–	0.009
4	1.000	0.915	0.992	–

Table 3.3 Description of the five cream cheeses

Cream cheese description	Product ID
Standard full fat cream cheese	1
Medium fat reduced cream cheese	2
Maximum fat reduced cream cheese	3
Prototype cream cheese	4
Prototype cream cheese + Butter Aroma	5

Table 3.4 Details on the six sensory descriptors of cream cheese

Domain	Descriptors	Abbreviation	Definition
	Resistance by hand	H-Resistance	Resistance during spreading with knife
Firmness	Firm by mouth	M-Firm	Hardness of sample in first press with tongue against palate
	Resistance by mouth	M-Resistance	Used force to dissolve food bolus
	Creaminess by mouth	M-Creaminess	Creaminess sensation in mouth
Creaminess	Cream flavor by mouth	M-Cream	Cream flavour assessed by mouth testing
	Cream aroma by nose	N-Cream	Cream aroma assessed by nose testing

It seems that product 1 and 2 are the best ones, especially for the first domain, while the remaining products look like not so different one each other.
We set up the ranking analysis problem as follows:

- type of design: multivariate randomized complete block design (five related samples, i.e. five multivariate populations to be ranked);
- domain analysis: yes (two domains);
- type of response variables: numerical continuous (six numerical responses);
- Ranking rule: "the higher the better";
- Combining function: Fisher;
- B—Number of permutations: 2000;
- Significance α-level: 0.05.

Fig. 3.3 Boxplot of sensory descriptor by product

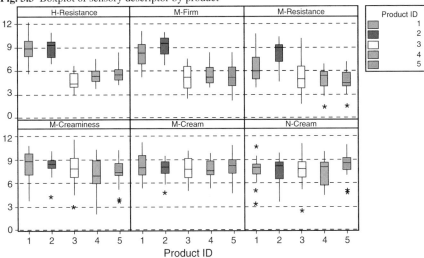

Fig. 3.4 Chart of sample means by sensory descriptor and product

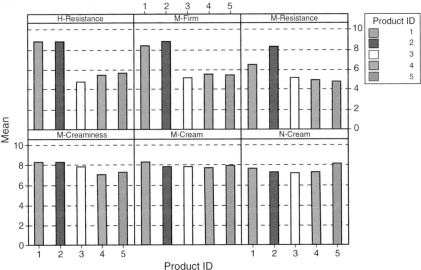

When applying the ranking method to wine quality data using directional parametric p-values performed via $B = 2000$ permutations we obtain results reported in
Tables 3.5 and 3.6, where directional multivariate permutation p-values in Table 3.1
have been adjusted by using the Bonferroni-Holm-Shaffer method [11].

Table 3.5 Firmness domain: Global ranking results ($\alpha = 1\%$) and related adjusted directional multivariate and directional (raw-unadjusted) univariate permutation p-values for the Cream Cheese sensory study

Firmness: all descriptors					
Global ranking	2	1	3	3	3
Product ID	1	2	3	4	5
1	–	0.925	0.005	0.003	0.003
2	0.005	–	0.003	0.003	0.002
3	1.000	1.000	–	1.000	1.000
4	1.000	1.000	0.369	–	1.000
5	1.000	1.000	0.231	1.000	–
H-Resistence					
Product ID	1	2	3	4	5
1	–	0.551	0.000	0.000	0.000
2	0.462	–	0.000	0.000	0.000
3	1.000	1.000	–	0.986	0.997
4	1.000	1.000	0.016	–	0.868
5	1.000	1.000	0.003	0.132	–
M-Firm					
Product ID	1	2	3	4	5
1	–	0.994	0.000	0.000	0.000
2	0.007	–	0.000	0.000	0.000
3	1.000	1.000	–	0.883	0.724
4	1.000	1.000	0.127	–	0.217
5	1.000	1.000	0.292	0.792	–
M-Resistence					
Product ID	1	2	3	4	5
1	–	1.000	0.012	0.000	0.000
2	0.000	–	0.000	0.000	0.000
3	0.989	1.000	–	0.315	0.155
4	1.000	1.000	0.698	–	0.272
5	1.000	1.000	0.850	0.744	–

It is worth noting the ranking results from the two domains are similar but somewhat different. Product 2 is the best product for firmness domain whose descriptors allow. in general to strongly discriminate among panellist assessment. In the case of creaminess domain there only very few significant differences and the best product seems to be Product 1.

Table 3.6 Creaminess domain: Global ranking results ($\alpha = 1\%$) and related adjusted directional multivariate and directional (raw-unadjusted) univariate permutation p-values for the Cream Cheese sensory study

Creaminess: all descriptors					
Global ranking	1	2	2	5	2
Product ID	1	2	3	4	5
1	–	0.597	0.123	0.005	0.093
2	1.000	–	1.000	0.360	0.450
3	1.000	1.000	–	0.648	0.861
4	1.000	1.000	1.000	–	0.951
5	1.000	0.753	0.720	0.360	–
M-Creaminess					
Product ID	1	2	3	4	5
1	–	0.453	0.163	0.001	0.006
2	0.559	–	0.171	0.006	0.007
3	0.849	0.840	–	0.046	0.087
4	1.000	0.995	0.958	–	0.733
5	0.997	0.994	0.920	0.280	–
M-Cream					
Product ID	1	2	3	4	5
1	–	0.043	0.018	0.019	0.078
2	0.963	–	0.543	0.471	0.686
3	0.984	0.478	–	0.419	0.674
4	0.983	0.548	0.597	–	0.734
5	0.930	0.327	0.342	0.282	–
N-Cream					
Product ID	1	2	3	4	5
1	–	0.204	0.182	0.179	0.898
2	0.807	–	0.523	0.504	0.984
3	0.830	0.493	–	0.491	0.987
4	0.832	0.504	0.521	–	0.992
5	0.107	0.017	0.016	0.010	–

3.3.3 Assessing Five Breads

Five different breads were baked in two replicates giving a total of ten samples. Eight different judges assessed the breads with respect to eleven different attributes in a fixed vocabulary profiling analysis. The data was kindly provided by Prof. Magni Martens (KVL, DK) and come from a student project in Sensory Science [3]. Sensory evaluations was performed in two replicates by a panel consisting of eight panellists. Figures 3.5 and 3.6 show the boxplot and the sample mean of sensory data by each sensory descriptor and product.

Fig. 3.5 Box plot of sensory attribute scores by bread. The empty dots represent the actual observed pannelist assessments

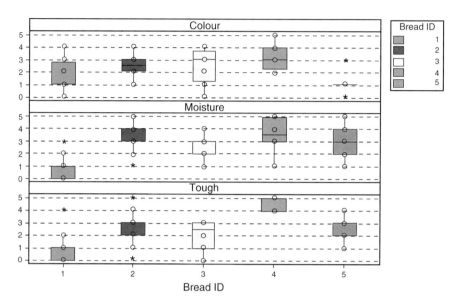

Fig. 3.6 Chart of sample means by sensory attribute score and bread

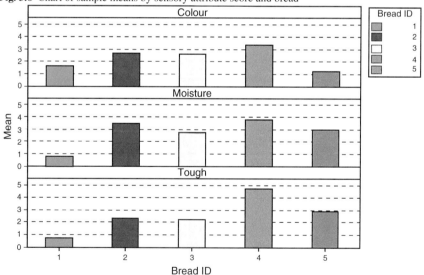

At a first sight, product 4 and 1 look as the best and the worst one, while the remaining products seems to differ only for some attributes. We set up the analysis ranking problem as follows:

- type of design: multivariate randomized complete block design (five related samples, i.e. five multivariate populations to be ranked);
- domain analysis: no;
- type of response variables: ordered categorical (three ordered categorical responses in a 0-5 rating scale);
- Ranking rule: "the higher the better";
- Combining function: Fisher;
- B—Number of permutations: 2000;
- Significance α-level: 0.01.

When applying the ranking analysis using directional permutation p-values ($\alpha = 5\%$), calculated via Anderson-Darling test statistic and performed with 2000 permutations and using the Fisher combining function, we obtain the following results (Table 3.7).

Table 3.7 Global ranking results ($\alpha = 1\%$) and related adjusted directional multivariate and directional (raw-unadjusted) univariate permutation p-values for the Five Breads sensory study

All attributes					
Global ranking	5	2	3	1	4
Product ID	1	2	3	4	5
1	–	1.000	1.000	1.000	1.000
2	0.005	–	0.762	1.000	0.005
3	0.003	1.000	–	1.000	0.114
4	0.003	0.003	0.002	–	0.003
5	0.003	1.000	0.446	0.999	–
Colour					
Product ID	1	2	3	4	5
1	–	0.997	0.986	1.000	0.151
2	0.006	–	0.490	0.991	0.000
3	0.027	0.643	–	0.990	0.001
4	0.000	0.026	0.023	–	0.000
5	0.932	1.000	1.000	1.000	–
Moisture					
Product ID	1	2	3	4·	5
1	–	1.000	1.000	1.000	1.000
2	0.000	–	0.046	0.824	0.148
3	0.000	0.985	–	0.998	0.852
4	0.000	0.258	0.006	–	0.028
5	0.000	0.928	0.268	0.988	–

(continued)

Table 3.7 (continued)

All attributes					
Tough					
Product ID	1	2	3	4	5
1	–	1.000	1.000	1.000	1.000
2	0.001	–	0.415	1.000	0.942
3	0.001	0.698	–	1.000	0.982
4	0.000	0.000	0.000	–	0.000
5	0.000	0.106	0.037	1.000	–

Ranking analysis confirms the significant much more better multivariate assessments of bread 4 over the other four breads. This result is mainly explained by the better performance of bread 4 in the assessments of attribute Tough. In fact, this is actually the attribute with the strongest significant differences (smallest p-values).

References

1. Arboretti Giancristofaro, R., Bonnini, S., Corain, L., Salmaso, L.: A permutation approach for ranking of multivariate populations. J. Multivar. Anal. **132**, 39–57 (2014)
2. Bertoluzzo, F., Pesarin, F., Salmaso, L.: On multi-sided permutation tests. Commun. Stat. Simul. Comput. **42**(6), 1380–1390 (2013)
3. Bro, R.: Multi-way Analysis in the Food Industry. Models, Algorithms, and Applications. 1998. Ph.D. thesis, University of Amsterdam (NL) & Royal Veterinary and Agricultural University (DK) (1998)
4. Bro, R., Qannari, El M., Kiers, H.A.L., Næs, T., Frøst M.B.: Multiway models for sensory profiling data. J. Chemom. **22**, 36–45 (2008)
5. Corain, L., Salmaso, L.: Improving power of multivariate combination-based permutation tests. Stat. Comput. **25**(2), 203–214 (2015)
6. Cortez, P., Cerdeira, A., Almeida, F., Matos, T., Reis, J.: Modeling wine preferences by data mining from physicochemical properties. Decis. Support Syst. **47**(4), 547–553 (2009)
7. Gupta, S.S., Panchapakesan, S.: Multiple Decision Procedures: Theory and Methodology of Selecting and Ranking Populations. SIAM-Society for Industrial and Applied Mathematics, Philadelphia (2002)
8. Hall, P., Miller, H.: Using the bootstrap to quantify the authority of an empirical ranking. Ann. Stat. **37**(6B), 3929–3959 (2009)

9. Meilgaard, M., Civille, G., Carr, B.: Sensory Evaluation Techniques, 4th edn. CRC Press, Boca Raton (2006)
10. Pesarin, F., Salmaso, L.: Permutation Tests for Complex Data: Theory, Applications and Software. Wiley, Chichester (2010)
11. Shaffer, J.P.: Modified sequentially rejective multiple test procedure. J. Am. Stat. Assoc. **81**, 826–831 (1986)

Chapter 4
Composite Indicators and Satisfaction Profiles

Evaluating the satisfaction about public services, organizations or products is very important in order to have a measure of their efficiency and effectiveness. For what concern the public service, in an interesting work Bird et al. [4] discuss about the importance of adopting the performance monitoring (PM) and performance indicators (PIs) also providing the steps to follow in order to reach them. Furthermore Marozzi [16] discusses the important role of the trust in public institutions and the importance of be able to assess it. In the years several authors debate the need of obtaining useful instruments for evaluating countries or geographical areas in their quality of life, job quality, living conditions etc. (see e.g. [5, 14, 10, 6, 12, 22, 26, 8]). In evaluating the quality of life of a country an important role is played by the satisfaction about its university system. The assessment of university satisfaction is widely discussed in literature (see for e.g. [7, 11, 15, 19]). Assessing satisfaction is not an easy issue since it is a complex phenomenon which is not directly observable and thus not directly measurable [15]. Most of the literature about evaluation of performances revolves around composite indicators which aggregate different dimensions or single indicators into a unique one. Indeed concepts as the impact of Government policies on public services, trust in public institutions, quality of work life, satisfaction on efficiency of university or school system etc. are multidimensional concepts which cannot be captured by a single indicator. The use of a synthesis indicator is not a trivial issue so that this topic has been widely discussed in literature and not always in a positive sense. In [13] and [25] we can find a summary of the main pros and cons of using composite indicators as discussed within the services of the European Commission whereas in [24] the main controversies on the use of statistical indices are discussed.

© The Author(s), under exclusive licence to Springer International Publishing AG,
part of Springer Nature 2018
R. Arboretti et al., *Parametric and Nonparametric Statistics for Sample Surveys
and Customer Satisfaction Data*, SpringerBriefs in Statistics,
https://doi.org/10.1007/978-3-319-91740-5_4

The construction of a composite indicator substantially consists of two steps: finding a data transformation T aiming at the comparability of different types of data and choosing a link function f for aggregating single indicators into a composite indicator. Thus a general composite indicator can be written as:

$$CI = f[T_1(x_1), T_2(x_2), \ldots, T_k(x_k)]$$

where x_i is the $i-th$ simple indicator, $i = 1, \ldots, k$.

For what concerns data transformation there exist several types of functions which lead to the comparability of data and they can be grouped in two main groups: linear transformations and non-linear transformations. Furthermore there are many ways to aggregate simple indicators with link functions. For an overview on the most widely used transformation and link functions see [1]. Other works with interesting discussion on aggregating functions are [20, 17, 23, 18, 27]. In practice most frequently adopted functions for pooling preferences ratings are the simple weighted or unweighted mean or summation, but when the distribution of the scores is not symmetric the mean is not valid, thus another way to summarize results is needed [2].

In what follow we introduce a suitable synthesis of a set of partial indicators following the nonparametric combination (NPC) of dependent tests methodology [21] and then we show the application of this composite indicator in analysing data from a students' satisfaction survey from the School of Engineering of the University of Padova.

4.1 NPC-Based Composite Indicator

The main purpose of NPC method is obtaining a single criterion for statistical units under study, which summarize many partial aspects. In order to formalize let us consider a k-dimensional variable $X = [X_1, \ldots, X_k]$ where the marginal variable $X_i, i = 1, \ldots, k$ assumes m_i ordered modalities $v_1, \ldots, v_{m_i}$ or discrete scores, $h = 1, \ldots, m_i, m_i \in N \setminus \{0\}, m_i > 1$. If X_i is a categorical variable, then a numerical transformation of modalities $v_1, \ldots, v_{m_i}$ into scores is needed. Large values of h correspond to higher satisfaction/quality rates. In order to simplify the notation, let us assume that $m_i = m$ for every i, but it is not necessary that all variables have the same number of modalities/scores. Let us suppose that these variables are given different (non-negative) degree of importance:

$$0 < w_i \leq 1, i = 1, \ldots, k.$$

Such weights are thought to reflect the different roles of the variables in representing indicators of the specific quality aspect under evaluation and are provided by responsible experts or from results of surveys previously carried out in the specific context.

4.1.1 Link Function

Suppose that N subjects give their judgement for k dependent variables each representing a specific aspect under evaluation. Thus the methodological problem we are going to face concerns obtaining a composite indicator which represents a global index of satisfaction for each subject starting from that k dependent variables. We introduce a set of minimum reasonable conditions related to variables $X_i, i = 1, \ldots, k$:

1. for each of the k informative variables a partial ordering criterion is well established, i.e. "large is better".
2. Regression relationships within the k informative variables are monotonic (increasing or decreasing).
3. The marginal distribution of each informative variables is non-degenerate.

It is worth to note that we do not need assuming continuity of $X_i, i = 1, \ldots, k$ so the probability of ex-equo can be positive. As link function we consider a real function ϕ from a class of Φ of real combining functions satisfying the following minimum properties:

1. ϕ must be continuous in all $2k$ arguments, in that small variations is any subset of arguments imply a small variation in the ϕ-index;
2. ϕ must be monotone non-decreasing with respect to each argument:

$$\phi(\ldots, X_i, \ldots; w_i, \ldots, w_k) \geq \phi(\ldots, X_i', \ldots, w_i, \ldots, w_k)$$
$$\text{if } 1 > X_i > X_i' > 0, i = 1, \ldots, k;$$

3. ϕ must be symmetric with respect to permutations of the arguments, in that if, for instance, $u_1, \ldots, u_k$, is any permutation of $1, \ldots, k$, then :

$$\phi(X_{u_1}, \ldots, X_{u_k}; w_{u_1}, \ldots, w_{u_k}) = \phi(X_1, \ldots, X_k; w_1, \ldots, w_k).$$

Property 1 is obvious. Property 2 means that if, for instance, two subjects have exactly the same values for all Xs except for the ith, then the one with $X_i > X_i'$ must have at least the same satisfaction ϕ-index assigned to it. Property 3 states that any combining function ϕ must be invariant with respect to the order in which informative variables are processed. As link function, the Fisher's combining function defined as: $\phi = -\sum_{i=1}^{k} w_i \times log(1 - \lambda_i)$, where $\lambda_i = (X_i + 0.5)/(m+1)$ are normalized scores defined in the open interval $[0, 1]$, satisfies the three described properties.

When dealing with assessment of satisfaction the Fisher's combining function appears more sensitive to assess higher satisfaction than to assess lower satisfaction, i.e. small differences in the lower satisfaction region seem to be identified with greater difficulty that those in the higher satisfaction region [9].

4.1.2 How to Compute the NPC-Based Composite Indicator Including Satisfaction Profiles

An aspect that should be considered when attempting to find a global index of satisfaction are *extreme* units, in the sense that the relevant question may not be the achievement of the absolute rank, but rather a more realistic expected one [4]. For this perspective we propose an extension of the NPC ranking method to the case of ordered categorical variables based on extreme satisfaction profiles [3]. Extreme satisfaction profiles are defined a priori on a hypothetical frequency distribution of variables $X_i, i = 1, \ldots, k$. Let us consider data X, where the rule "large is better" holds for all variables. Observed values for the k variables are denoted by $x_{ji}, i = 1, \ldots, k; j = 1, \ldots, N$. Examples of extreme satisfaction profiles are given below.

The *strong* satisfaction profile is defined as follows.

Maximum satisfaction is obtained when all subjects have the highest value of satisfaction for all variables:

$$f_{hi} = \begin{cases} 1 & \text{for } h = m \\ 0 & \text{otherwise} \end{cases} \forall i, i = 1, \ldots, k$$

where f_{hi} are the relative frequencies of categories $h, h = 1, \ldots, m$, for variable $X_i, i = 1, \ldots, k$.

Minimum satisfaction is obtained when subjects have the smallest value of satisfaction with relative frequencies varying across variables:

$$f_{hi} = \begin{cases} 1 & \text{for } h = 1 \\ 0 & \text{otherwise} \end{cases} \forall i, i = 1, \ldots, k$$

The *weak* satisfaction profile is defined as follows.

Maximum satisfaction is obtained when all subjects have the highest value of satisfaction for all variables:

$$f_{hi} = \begin{cases} u_i & \text{for } h = m \\ u_{hi} & \text{otherwise, where } \sum_{h=1}^{m-1} u_{hi} = (1 - u_i) \ i = 1, \ldots, k \end{cases}$$

Minimum satisfaction is obtained when subjects have the smallest value of satisfaction with relative frequencies varying across the variables:

$$f_{hi} = \begin{cases} l_i & \text{for } h = 1 \\ l_{hi} & \text{otherwise, where } \sum_{h=2}^{m} l_{hi} = (1 - l_i) \ i = 1, \ldots, k \end{cases}$$

where u_i and l_i represent realistic achievable targets that can be fixed observing past experience or motivational targets established by managers or organizers in the strategic and business planning.

In order to include the extreme satisfaction profiles in the analysis, we transform original values $h, h = 1, \ldots, m$ as follow: separate the values of h corresponding to a judgement of satisfaction, say the last $t, 1 \leq t \leq m$, from those corresponding to judgements of dissatisfaction, i.e. $(m - t)$. For the last t values of h corresponding to judgements of dissatisfaction, the transformed values of h are defined as:

$$h + f_{hi} \times 0.5 h = m - t + 1, \ldots, m; i = 1, \ldots, k.$$

For the first $(m - t)$ values of h corresponding to judgements of dissatisfaction, the transformed values of h are defined as:

$$h + (1 - f_{hi}) \times 0.5 h = 1, \ldots, m - t; i = 1, \ldots, k.$$

This transformation is equivalent to the assignment to the original values $h, h = 1, \ldots, m$, of additive degrees of importance which depend on relative frequencies f_{ih} and which increase the original values h up to $h + 0.5$. The limit 0.5 is fixed in such a way that the increase in the original score h, positively (negatively) related to the fraction of evaluators who choose the corresponding judgement, is less than one; hence the transformation of h is less than $h + 1$. Let us suppose, for example, that $h = 1, 2, 3, 4$ and values 3 and 4 correspond to judgements of satisfaction. By applying the above transformation, the value of 3 tends towards the upper value for, which represents higher satisfaction, when f_{i3} increases. On the contrary, the value of 1 tends towards 2 (less dissatisfaction) when f_{i1} decreases. The transformation of values $h, h = 1, \ldots, m$ weighted by relative frequencies f_{ih}, is applied to observed values $x_{ji}, i = 1, \ldots, k; j = 1, \ldots, N$. For the last t values of h corresponding to a judgement of satisfaction, the transformed values of x_{ji} are defined as:

$$z_{ji} = x_{ji} + \sum_{h=m-t+1}^{m} I_h(x_{ji}) \times f_{ih} \times 0.5, i = 1, \ldots, k; j = 1, \ldots, N,$$

where

$$I_h(x_j i) = \begin{cases} 1 & \text{if } x_{ji} = h \\ 0 & \text{if } x_{ji} \neq h \end{cases}$$

For the first $(m - t)$ values of h corresponding to judgments of dissatisfaction, the transformed values of x_{ji} are defined as:

$$z_{ji} = x_{ji} + \sum_{h=1}^{m-t} I_h(x_{ji}) \times (1 - f_{ih}) \times 0.5, i = 1, \ldots, k; j = 1, \ldots, N.$$

In this setting, we can consider the following transformations:

$$\lambda_{ji} = \frac{(z_{ji} - z_{i\min}) + 0.5}{(z_{i\max} - z_{i\min}) + 1}, i = 1, \ldots, k; j = 1, \ldots, N$$

with $z_{i\min}$ and $z_{i\max}$ obtained according to an extreme satisfaction profile. If we consider the strong satisfaction profile, we have:

$$z_{i\min} = x_{ji} + \sum_{h=1}^{m-t} I_h(x_{ji}) \times (1 - f_{ih}) \times 0.5 = 1,$$

where $f_{ih} = 1$ and $x_{ji} = h = 1, i = 1, \ldots, k,$

$$z_{i\max} = x_{ji} + \sum_{h=m-t+1}^{m} I_h(x_{ji}) \times 0.5 = m + 0.5,$$

where $f_{ih} = 1$ and $x_{ji} = h = m, i = 1, \ldots, k$. If we consider a weak satisfaction profile, for example with $u = 0.7$ and $l = 1$, we have:

$$z_{i\min} = x_{ji} + \sum_{h=1}^{m-t} I_h(x_{ji}) \times (1 - f_{ih}) \times 0.5 = 1,$$

where $f_{ih} = 1$ and $x_{ji} = h = 1, i = 1, \ldots, k,$

$$z_{i\max} = x_{ji} + \sum_{h=m-t+1}^{m} I_h(x_{ji}) \times 0.5 = m + 0.35,$$

where $f_{ih} = 0.7$ and $x_{ji} = h = m, i = 1, \ldots, k.$

Note that $z_{i\max}$ represents the preferred value for each variable, and it is obtained when satisfaction is at its highest level according to the extreme satisfaction profile; $z_{i\min}$ represents the worst value, and it is obtained when satisfaction is at its lowest level according to the extreme satisfaction profile. Scores $\lambda_{ji}, i = 1, \ldots, m, j = 1, \ldots, N$ are one-to-one increasingly related to values x_{ji}, z_{ji}. In order to synthesize the k-partial rankings based on scores $\lambda_{ji}, i = 1, \ldots, m, j = 1, \ldots, N$ using NPC ranking method, we use a combining function ϕ:

$$T_j = \phi(\lambda_{j1}, \ldots, \lambda_{jk}; w_1, \ldots, w_k), j = 1, \ldots, N.$$

In order for the global index to vary in the interval $[0, 1]$ we put:

$$S_j = \frac{T_j - T_{\min}}{T_{\max} - T_{\min}}, j = 1, \ldots, N,$$

where

$$T_{\min} = \phi(\lambda_{1\min}, \ldots, \lambda_{k\min}; w_1, \ldots, w_k),$$
$$T_{\max} = \phi(\lambda_{1\max}, \ldots, \lambda_{k\max}; w_1, \ldots, w_k),$$

and $\lambda_{i\min}$ and $\lambda_{i\max}$ are obtained according to the extreme satisfaction profiles:

$$\lambda_{i\min} = \frac{(z_{i\min} - z_{i\min}) + 0.5}{(z_{i\max} - z_{i\min}) + 1}, i = 1, \ldots, k,$$

$$\lambda_{i\max} = \frac{(z_{i\max} - z_{i\min}) + 0.5}{(z_{i\max} - z_{i\min}) + 1}, i = 1, \ldots, k,$$

Note that value T_{min} represents the *unpreferred value* of the satisfaction index since it is calculated from $(\lambda_{i\,min}, \ldots, \lambda_{k\,min})$, while T_{max} represents the *preferred value* since it is calculated from $(\lambda_{i\,max}, \ldots, \lambda_{k\,max})$. T_{min} and T_{max} are reference values to evaluate the distance of the observed satisfaction values from the situation of highest satisfaction defined according to the extreme satisfaction profile.

4.2 A Students' Satisfaction Survey

At the end of each teaching course, the students of the School of Engineering of the University of Padova in Italy are required to complete a questionnaire on their satisfaction about it. The questionnaire covers different aspects of satisfaction such as organizational aspects, teaching activities and infrastructures. Finally students were asked to give a judgement on their *overall* satisfaction. Questions of different domains are shown in Table 4.1.

Table 4.1 Questions from each domain of satisfaction

Domain	Questions
Organizational Aspects	**D01**. At the beginning of the course the aims and the contents were clearly presented?
	D02. The examination procedures were clearly defined?
	D03. The times of teaching activities were complied with?
	D04. The recommended course material was appropriate?
Teaching Activities	**D05**. The teacher encouraged/motivated the interest in the subject?
	D06. The teacher set out the topics clearly?
	D07. The professor was during his office hours for clarifications and explanations?
	D08. Workshops, tutorial and seminars, if any, were appropriate?
Overall Satisfaction	**D09**. How much are you satisfied with the development of the course on the whole?

Answers consist of scores in Likert scale 1–10 intended as "greater is better". We analyze data from this survey adopting the NPC-based composite indicator described in previous section. It is common in practice to consider the simple mean of the answers to the question related to the *overall* satisfaction (D09 of Table 4.1) as indicator of global satisfaction about a teaching course. In the following sections we aim at showing the advantages that NPC-based composite indicator brings in understanding the real students' satisfaction.

4.2.1 Satisfaction Profiles

In Sect. 4.1.2 we discuss the importance of considering different satisfaction profiles. In order to better understand the importance of this feature let us consider a
very simple example. A teaching course of the first year of Management Engineering is attended by 300 students. Suppose we want to assess the quality of the course
on the basis of the satisfaction about room (e.g. enough seats, good acoustics etc.)
and the satisfaction about the quality of teaching (e.g. if teacher explains well). Here
we can think to set two different benchmarks of satisfaction for the two aspects because different is the expected satisfaction from them. Indeed in a room with a lot
of students is unlikely to expect the highest satisfaction from all students about the
room, this is because for example, best seats are given early. On the contrary, we expect that all students are highly satisfied by the teacher. For these reasons it is reasonable to think that the maximum expected satisfaction about room is reached when
at least the 60% of students have the highest satisfaction, whereas the maximum
satisfaction expected for teaching is that the 100% of the students have the highest satisfaction. Figure 4.1 shows how results and related conclusions, can change
adopting different satisfaction profiles. Since the NPC composite indicator $\in [0, 1]$
let us suppose of considering the point 0.5 as a threshold of sufficient satisfaction.
We can see that considering a strong satisfaction profile, where the benchmark of
maximum satisfaction is that all students have the highest satisfaction for all partial
aspects, we obtain a composite indicator with a median around 0.40, that is under
the threshold of sufficient satisfaction. Thus we conclude that this course globally
does not satisfy sufficiently, students who attend it. On the other hand, when we
consider a weak satisfaction profile assigning different benchmarks of satisfaction
for different aspects (in line with the actual expectations) we obtain an indicator
with a median of 0.52 that is over the threshold of sufficient satisfaction. In this case
the same teaching course globally, sufficiently satisfy students who attend it.

4.2.2 Taking into Account All Partial Aspects

We also investigate the impact of single aspects of satisfaction both on overall satisfaction intended as the answers to D09 and on the NPC composite indicator, by
means of a multiple linear regression model (other model could be also adopted as
latent classes, multilevel models, etc.). We would expect that the overall satisfaction
is influenced simultaneously by all partial aspects. Actually when we are asked to
express a judgement about something a complex mechanism is activated in our mind
and it is common to be guided only by a particular aspect or few aspects. Indeed analyzing data from students' satisfaction we found that the overall satisfaction seems
guided by the satisfaction on how much teacher motivates the interest in the sub-

Fig. 4.1 Distribution of the NPC-based composite indicator using different satisfaction profiles. Red dashed lines represent the point of sufficient satisfaction; black dashed lines represent the median of the composite indicator

Weak Satisfaction Profile

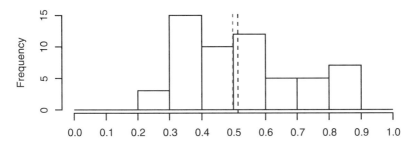

Strong Satisfaction Profile

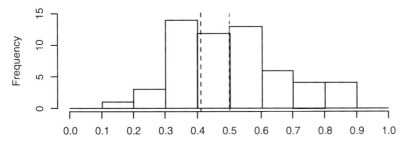

ject (D05). Table 4.2 shows a representative extract of the results after adopting for all teaching courses a regression model putting in relation the overall satisfaction as indicated in D09 with partial aspects D01-D06. Since answers to D07-D09 presented a lot of missing values, we exclude them from this analysis. As we can see satisfaction about D05 is present for each teaching course whereas other aspects do not always influence the satisfaction. It is more evident in Fig. 4.2 where we report the histogram showing the percentage of time when each partial aspect is resulted significant (at a significance level $\alpha = 0.05$) in the multivariate regression model when putted in relation with D05 and NPC composite indicator. We can see that the aspect related to the motivation of the teacher (D05) in the 80% of times impacts on the overall satisfaction as intended by D09, followed by aspect related to the teaching material (D04) and aspect related to teacher explanation (D06) both with about 40%. Whereas the NPC composite indicator obviously (for its construction) is influenced by all partial aspects (the distribution is almost uniform).

Table 4.2 Extract of significant partial aspects in the regression model which put in relation the overall satisfaction (D09) with partial aspects in the questionnaire

Teaching Course ID	D01	D02	D03	D04	D05	D06
TC_ID1			*		*	
TC_ID2		*			*	
TC_ID3	*			*	*	
TC_ID4					*	
TC_ID5					*	*
TC_ID6	*				*	
TC_ID7		*			*	*
TC_ID8		*		*	*	
TC_ID9					*	
...		...	...	...	...	...

The star means that the aspect is significative at a significance level $\alpha = 0.05$

Fig. 4.2 Percentage of time when each partial aspect results significant ($\alpha = 0.05$) in determining the overall satisfaction

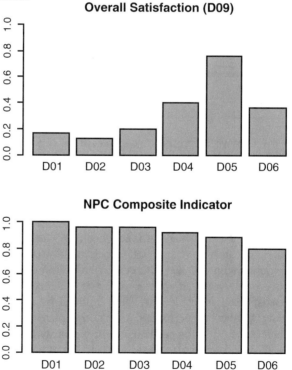

In order to emphasize how the NPC-based composite indicator well explains the satisfaction structure of the respondents we show for a teaching course the distribution of the scores of each partial aspects, of the overall satisfaction and of the composite indicator (see Fig. 4.3). What we can see is that for example for this teaching course, the distribution of the scores of the overall satisfaction (D9) is very much similar to that related to teaching explanation (D6) confirming that the overall satisfaction seems to be guided by a particular aspect. On the other hand the scores of the composite indicator, converted on a scale 1–10, does not follow a specific aspect but it mediate all partial aspects.

Fig. 4.3 Example of distribution of the scores of a teaching course for each partial aspect, overall satisfaction and composite indicator

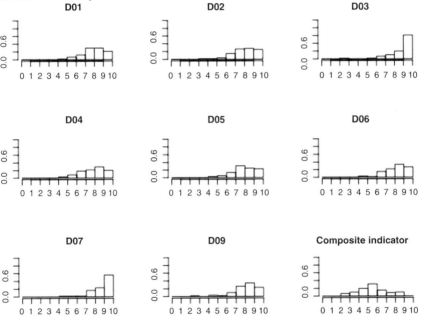

References

1. Aiello, F., Attanasio, M.: How to transform a batch of simple indicators to make up a unique one? Atti della XLII Societa italiana di Statistica, Sessioni Plenarie e Specializzate, CLEUP, Padova, pp. 327–338 (2004)
2. Arboretti Giancristofaro, R., Marozzi, M., Salmaso, L.: Nonparametric pooling and testing of preference ratings for full-profile conjoint analysis experiments. J. Mod. Appl. Stat. Methods **4**(2), 353–628 (2005)

3. Arboretti Giancristofaro, R., Bonnini, S., Salmaso, L.: Employment status and education/employment relationship of PhD graduates from the University of Ferrara. J. Appl. Stat. **36**(12), 1329–1344 (2009)
4. Bird, S.M., Cox, D., Farawell, V.T., Goldstain, H., Holt, T., Smith, P.C.: Performance indicators: good, bad and ugly. J. R. Stat. Soc. Ser. A Stat. Soc. **168**, 1–27 (2005)
5. Boccuzzo, G., Gianecchini, M.: Measuring young graduates' job quality through a composite indicator. Soc. Indic. Res. **122**(2), 453–478 (2015)
6. Coromaldi, M., Zoli, M.: Deriving multidimensional poverty indicators: methodological issues and an empirical analysis for Italy. Soc. Indic. Res. **107**(1), 37–54 (2012)
7. Del Carmen Bas, M., Tarantola, S., Carot, J.M., Conchado, A.: Sensitivity analysis: a necessary ingredient for measuring the quality of a teaching activity index. Soc. Indic. Res. **131**(3), 931–946 (2017)
8. Fayers, P.M., Hand, D.J.: Casual variables, indicator variables and measurement scales: an example from quality of life. J. R. Stat. Soc. Ser. A Stat. Soc. **165**(2), 233–253 (2002)
9. Finos, L., Salmaso, L.: Nonparametric multi-focus analysis for categorical variables. Commun. Stat. Theory Methods **33**(8), 1931–1941 (2004)
10. Giambona, F., Vassallo, E.: Composite indicators of social inclusion for European countries. Soc. Indic. Res. **116**(1), 269–293 (2014)
11. Gnaldi, M., Ranalli, M.G.: Measuring university performance by means of composite indicators: a robustness analysis of the composite measure used for benchmark of Italian universities. Soc. Indic. Res. **129**(2), 659–675 (2016)
12. Hoskins, B.L., Mascherini, M.: Measuring active citizenship through the development of a composite indicator. Soc. Indic. Res. **90**(3), 459–488 (2009)
13. Joint Research Center - European Commission: Handbook on Constructing Composite Indicators: Methodology and User Guide. OECD Publishing, Paris (2008)
14. Krishnan, V.: Development of a multidimensional living conditions index (LCI). Soc. Indic. Res. **120**(2), 455–481 (2015)
15. Marozzi, M.: A composite indicator dimension reduction procedure with application to university student satisfaction. Stat. Neerl. **63**(3), 258–268 (2009)
16. Marozzi, M.: Construction, dimension reduction and uncertainty analysis of an index of trust in public institutions. Qual. Quant. **48**(2), 939–953 (2014)
17. Munda, G.: Choosing aggregation rules for composite indicators. Soc. Indic. Res. **109**(3), 337–354 (2012)
18. Munda, G., Nardo, M., Saisana, M., Srebotnjak, T.: Measuring uncertainties in composite indicators of sustainability. Int. J. Environ. Technol. Manag. **11**, 7–26 (2009)
19. Murias, P., De Miguel, J.C., Rodríguez, D.: A composite indicator for university quality assessment: the case of Spanish higher education system. Soc. Indic. Res. **89**(1), 129–146 (2008)
20. Paruolo, P., Saisana M., Saltelli, A.: Rating and rankings: voodoo or science? J. R. Statist. Soc. A **176**, 609–634 (2013)

21. Pesarin, F., Salmaso, L.: Permutation tests for complex data, theory, applications and software. Wiley, Chichester (2010)
22. Royuela, V., López-Tamayo, J., Suriñach, J.: Results of a quality of work life index in Spain. A comparison of survey results and aggregate social indicators. Soc. Indic. Res. **90**(2), 225–241 (2009)
23. Saisana, M., d'Hombres, B., Saltelli, A., Rickety numbers: volatility of university rankings and policy implications. Res. Policy **40**, 165–177 (2011)
24. Saltelli, A.: Composite indicators between analysis and advocacy. Soc. Indic. Res. **81**(1), 65–77 (2007)
25. Saltelli, A., Tarantola, S.: On the relative importance of input factors in mathematical models: safety assessment for nuclear waste disposal. J. Am. Stat. Assoc.
26. Somarriba, N., Pena, B.: Synthetic indicators of quality of life in Europe. Soc. Indic. Res. **94**(1), 115–133 (2009)
27. Tarabusi, C.E., Guarini, G.: An unbalance adjustment method for development indicators. Soc. Indic. Res. **112**, 19–45 (2013)

Chapter 5
Analyzing Survey Data Using Multivariate Rank-Based Inference

Data from customer satisfaction surveys are multivariate—there are several questions resulting in as many endpoints. Furthermore, survey data typically don't fit into simple parametric models. Indeed, the endpoints or response variables may be measured on different types of scales (metric, ordinal, binary). For these two reasons, one requires multivariate inference methods, and specifically methods that can deal with a mix of response variable types. Additionally, it would be advantageous if the procedures also performed well for small to moderate numbers of respondents, as not every survey can afford to obtain responses from hundreds of participants.

Appropriate inference procedures fulfilling all these requirements for multivariate data are rare. However, they do exist within the framework of the multivariate, robust rank-based approach developed in the last decade through the publications [1, 2, 4, 3, 11, 10, 12, 15], and implemented in the R package npmv [9, 19]. This approach constitutes a nonparametric extension of parametric multivariate analysis of variance (MANOVA), and it doesn't suffer from the severe limitations of those classical approaches. Another set of appropriate inference procedures for multivariate non-normal data is described in the book by Pesarin and Salmaso [18] and references cited therein. In this chapter, we will focus on the former approach, the latter is described in other chapters of the book.

5.1 Why Should Classical Parametric MANOVA Methods *Not* Be Used for Categorical Responses?

Recall that the classical MANOVA model assumes that observation vectors are realizations of multivariate normal random variables. Furthermore, the classical model assumes that the covariances structure between the different outcomes doesn't differ between groups. That is, no matter which group of respondents is considered, the endpoints have the same variances and covariances. This latter assumption is called *homoscedasticity assumption*.

To be mathematically precise, the classical model assumes the following.

$$(X_{ij}^{(1)}, \ldots, X_{ij}^{(k)})' \sim N_k(\mu_i, \Sigma), \ i = 1, \ldots, a; \ j = 1, \ldots, n_i,$$

where the X_{ij} are independent random vectors of dimension k, following a multivariate normal distribution. The index i stands for a group or sub-population (e.g., male vs. female respondents), j represents a particular experimental unit or subject within the group i (e.g., male respondents Adam, Bobby and Charlie, as well as female respondents Debby and Evelyn), and there are k variables or endpoints being measured on each subject (e.g., answers to k different survey questions). The k-dimensional vector of expected values, μ_i, may differ between groups. Its equality across groups is a typical null hypothesis in parametric models. The covariance matrix Σ, however, is assumed to be equal across groups, when the classical model is postulated.

Are the classical assumptions of multivariate normality and homoscedasticity fulfilled in practice? Or can they at least be checked in a meaningful way?

As an example, consider a survey with male and female respondents who are asked about the quality of instruction at a sports school using two different questions with ordinal response scale. First of all, it is clear that ordinal responses can by no means be modelled through multivariate normal random variables. Even for metric data, reliably checking the assumption of multivariate normality is practically impossible. Assessing univariate normality is often a difficult task already, in particular for smaller data sets. However, for this illustrative bivariate example with two ordinal endpoints, the use of classical MANOVA is clearly out of question.

ANOVA and MANOVA are based on comparing averages of responses, and taking their differences, but neither averages nor differences are meaningful concepts for ordinal data. Differences between answers don't have the clearly defined meaning that they have for metric data. For example, the difference between categories "very good" and "good" may not be perceived as being of equal size as the difference between "good" and "average" or "satisfactory". If differences between responses don't have a uniquely defined meaning in a particular situation, then also the plethora of statistical methods developed for metric responses is not appropriate for this situation. A few decades ago, appropriate nonparametric methods for the analysis of multivariate ordinal data didn't exist yet. Arguably, it is largely for this reason that some researchers advocated use of the clearly inappropriate (M)ANOVA methods in order to have at least some inferential techniques available to tackle their data. This surely didn't contribute to research reproducibility, which has become a major issue of concern in several fields, and we have to strongly advise against this strategy.

Apart from the obvious violation of the normality assumption in this example, the assumption of equal covariance matrices is also a severe limitation in practice. It would mean that, for example, male and female respondents would show the same degree of variation in answering the two questions, and the same correlation between the two responses. These are some rather strong assumptions. Why would anyone who has no information regarding equal means, and thus wanting to test their equality, believe in equal variances and correlations a priori? Even worse, violation of this so-called homoscedasticity assumption has rather detrimental effects on the performance of classical MANOVA methods. Just as the two-sample t-test with equal variance assumption performs poorly when variances are not equal and samples not balanced, its extensions to ANOVA and MANOVA suffer from the same problems. They may become very conservative (large type II error probability) or very liberal (large type I error probability), depending on the configuration of sample sizes and underlying variances.

Concluding, there are obvious situations where classical MANOVA doesn't make sense for the data at hand because the response variables are not metric. However, even for metric data, there is no guarantee that the actual error rates of classical MANOVA procedures are close to the nominal error rates when the strong assumptions of multivariate normality and homoscedasticity are not fulfilled. And, if the error rates are not reliable, those methods should not be used.

5.2 How Does the Nonparametric Multivariate Model Look Like?

A fully nonparametric model for the multivariate observation vectors can be formulated as follows.

$$(X_{ij}^{(1)}, \ldots, X_{ij}^{(k)})' \sim F_i, \ i = 1, \ldots, a; \ j = 1, \ldots, n_i,$$

where the X_{ij} are again independent random vectors of dimension k, and there is a total of $N = \sum_{i=1}^{a} n_i$ respondents or response vectors.

The details of this nonparametric model formula look a bit simpler compared to the parametric model above. In fact, there are no assumptions regarding any particular distribution, such as a multivariate normal distribution. Instead, the model simply states that the observations from group i follow some multivariate distribution that is denoted as F_i. Each group may have a different multivariate distribution. It may be multivariate normal in one group, and multivariate exponential in another. Or, it may even be some discrete or absolutely continuous distribution without name. In fact, one endpoint may be ordinal, another endpoint may be quantitative (metric), and a third endpoint may be binary. Clearly, the data-generating model could not be more general.

Also, there are no assumptions on specific covariance matrix structures. In general, the covariance structure between the endpoints may be different for each group. However, the generality of this model comes along with a rather stringent null hypothesis. For inference in the sense of hypothesis testing, the null hypothesis is formulated in terms of equality of the multivariate distributions, namely

$H_0^F : F_1 = \cdots = F_a$. Thus, under this *global null hypothesis*, all multivariate distributions are equal. This implies that under null hypothesis, the means (if well-defined), the medians, but also the dispersions are equal in each group. A bit later, when we discuss the details on simultaneously testing subsets of the endpoints and of the groups, this stringent null hypothesis will be softened somewhat.

Finally, the approach explained in this chapter does not require that group effects point into the same direction for each variable, an assumption made in the test by [17] that is widely cited in the medical literature. For example, an improvement in a treatment, or a more favorable view of one sub-population, may in reality correspond to increasing values of one outcome, and decreasing values of another, while even other response variables may be completely unaffected. Such a situation would present a challenge for methods designed similarly as the 'O"Brien test. However, in order to apply the methods presented here, it is not at all of importance. That is, it is also *not* necessary to transform the variable scales in such a way that a favorable outcome always corresponds to larger values. In fact, no transformation at all is needed in order to perform inference using the nonparametric multivariate methods described here.

5.3 How Can Inference Be Performed for the *Global* Null Hypothesis?

The global null hypothesis states that all k-variate distributions are equal for all a groups that are being compared, namely $H_0^F : F_1 = \cdots = F_a$. Valid inference for this global null hypothesis can be performed using the following steps. All the steps described here in detail can conveniently be performed automatically using the R package npmv. However, we think it is instructive and helpful in a knowledgeable interpretation of the results to know what is implemented in the software, and in part also *why*. We will illustrate the correct use of the R package in the subsequent section.

1. **Ranking.**
   Replace the observations for each endpoint by their mid-ranks. That is, for each response variable separately, the observations (across all groups) are ranked from smallest (rank 1) to largest (rank N). This is called variable-wise ranking. In case of ties, mid-ranks are recommended. Table 5.1 illustrates the rank calculation for a simple example with three variables ($k = 3$) and two groups ($a = 2$) of size $n_1 = 4$ and $n_2 = 3$, respectively. The example is deliberately chosen to include outcomes measured on three different scales, namely a metric, an ordinal, and a binary (dichotomous) variable. All three scales typically appear in surveys, and the approach presented here is general enough to allow for valid inference in the presence of even a mixture of the three. As long as variable-wise ranking is possible, the respective response variables can be included. The resulting ranks are assembled in the matrix **R** given in Table 5.2. This matrix has k rows, correspond-

ing to the k endpoints, as well as N columns. Each column represents the ranks obtained for one specific respondent. The blocks indicate different groups of respondents (for example, males vs. females, or different nationalities, or different types of instruction experienced).

Table 5.1 Illustration of variable-wise rank calculations for $k = 3$ variables and $a = 2$ groups

Original observations							(Mid-)ranks						
Group 1				Group 2			Group 1				Group 2		
$X_{11}^{(1)}$	$X_{12}^{(1)}$	$X_{13}^{(1)}$	$X_{14}^{(1)}$	$X_{21}^{(1)}$	$X_{22}^{(1)}$	$X_{23}^{(1)}$	$R_{11}^{(1)}$	$R_{12}^{(1)}$	$R_{13}^{(1)}$	$R_{14}^{(1)}$	$R_{21}^{(1)}$	$R_{22}^{(1)}$	$R_{23}^{(1)}$
3	2	1	4	2	2	1	6	4	1.5	7	4	4	1.5
$X_{11}^{(2)}$	$X_{12}^{(2)}$	$X_{13}^{(2)}$	$X_{14}^{(2)}$	$X_{21}^{(2)}$	$X_{22}^{(2)}$	$X_{23}^{(2)}$	$R_{11}^{(2)}$	$R_{12}^{(2)}$	$R_{13}^{(2)}$	$R_{14}^{(2)}$	$R_{21}^{(2)}$	$R_{22}^{(2)}$	$R_{23}^{(2)}$
-0.03	0.64	0.75	-0.80	0.97	0.24	-1.64	3	5	6	2	7	4	1
$X_{11}^{(3)}$	$X_{12}^{(3)}$	$X_{13}^{(3)}$	$X_{14}^{(3)}$	$X_{21}^{(3)}$	$X_{22}^{(3)}$	$X_{23}^{(3)}$	$R_{11}^{(3)}$	$R_{12}^{(3)}$	$R_{13}^{(3)}$	$R_{14}^{(3)}$	$R_{21}^{(3)}$	$R_{22}^{(3)}$	$R_{23}^{(3)}$
0	0	0	1	1	0	1	2.5	2.5	2.5	6	6	2.5	6

The two groups have $n_1 = 4$ and $n_2 = 3$ respondents, respectively. The first outcome is measured on an ordinal scale with categories coded as numbers 1–5. The second outcome is metric, rounded to two digits, while the third endpoint is binary. In case of ties, midranks are calculated. The inferential methods presented in the text allow for such a mixture of outcome scales

Table 5.2 Matrix $\mathbf{R}$ of variable-wise mid-ranks

Group 1			Group 2			...	Group a		
$R_{11}^{(1)}$	$R_{12}^{(1)}$	$\cdots$ $R_{1n_1}^{(1)}$	$R_{21}^{(1)}$	$R_{22}^{(1)}$	$\cdots$ $R_{2n_2}^{(1)}$	$\cdots$	$R_{a1}^{(1)}$	$R_{a2}^{(1)}$	$\cdots$ $R_{a,n_a}^{(1)}$
$R_{11}^{(2)}$	$R_{12}^{(2)}$	$\cdots$ $R_{1n_1}^{(2)}$	$R_{21}^{(2)}$	$R_{22}^{(2)}$	$\cdots$ $R_{2n_2}^{(2)}$	$\cdots$	$R_{a1}^{(2)}$	$R_{a2}^{(2)}$	$\cdots$ $R_{a,n_a}^{(2)}$
$\cdots$			$\cdots$			$\cdots$	$\cdots$		
$R_{11}^{(k)}$	$R_{12}^{(k)}$	$\cdots$ $R_{1n_1}^{(k)}$	$R_{21}^{(k)}$	$R_{22}^{(k)}$	$\cdots$ $R_{2n_2}^{(k)}$	$\cdots$	$R_{a1}^{(k)}$	$R_{a2}^{(k)}$	$\cdots$ $R_{a,n_a}^{(k)}$

Each row corresponds to one of the k endpoints, and each column corresponds to one of the N respondents. The N respondents are divided up into a groups with possibly differing sample sizes n_i, $i = 1, \ldots, a$. The jth column in the ith group will be denoted as $\mathbf{R}_{ij}$. The average of all n_i columns in the ith group is $\bar{\mathbf{R}}_{i.}$, and $\bar{\mathbf{R}}_{..}$ denotes the average of all N columns

2. **Estimating matrices representing sums of squares and cross-products.**
   Calculate the between- and within-group covariance matrices of these rank data. This is similar to calculating the numerator and denominator sums of squares in analysis of variance. The details are somewhat technical. We will denote the between-group covariance matrix by $\mathbf{H}$, and the within-group covariance matrix by $\mathbf{G}$. There is another subtle detail in that for each of these matrices $\mathbf{H}$ and $\mathbf{G}$, different versions exist, depending on whether weighted or unweighted means shall be used in the sums of squares calculations. However, the practical performance differences between these choices are actually minor. In Eqs. (5.1) and (5.2), we only give the formulas for weighted means and refer to the appendix of Ellis et al. [9], as well as the technical articles mentioned above, for further details and alternative definitions.

$$\mathbf{H} = \frac{1}{a-1} \sum_{i=1}^{a} n_i (\bar{\mathbf{R}}_{i.} - \bar{\mathbf{R}}_{..})(\bar{\mathbf{R}}_{i.} - \bar{\mathbf{R}}_{..})^\top = \frac{1}{a-1} \mathbf{R} \left(\bigoplus_{i=1}^{a} \frac{1}{n_i} J_{n_i} - \frac{1}{N} J_N \right) \mathbf{R}^\top ,$$

$$(5.1)$$

$$\mathbf{G} = \frac{1}{N-a} \sum_{i=1}^{a} \sum_{j=1}^{n_i} (\mathbf{R}_{ij} - \bar{\mathbf{R}}_{i.})(\mathbf{R}_{ij} - \bar{\mathbf{R}}_{i.})^\top = \frac{1}{N-a} \mathbf{R} \left(\bigoplus_{i=1}^{a} P_{n_i} \right) \mathbf{R}^\top . \qquad (5.2)$$

The $(k \times N)$-dimensional matrix $\mathbf{R}$, and the k-dimensional vectors $\mathbf{R}_{ij}$, $\bar{\mathbf{R}}_{i.}$, and $\bar{\mathbf{R}}_{..}$ have been introduced in Table 5.2. The matrix J is a square matrix of ones (dimension indicated by the respective subscript), and P_{n_i} is a square matrix with diagonal elements $(1 - 1/d)$, while all other elements of this matrix are equal to $(-1/d)$

The rightmost parts of Eqs. (5.1) and (5.2) define the matrices $\mathbf{H}$ and $\mathbf{G}$ in the mathematically more elegant way which also allows for a straightforward implementation in statistical software. The left part uses the traditional summation notation that can be found in classical textbooks on analysis of variance (ANOVA) and related methods. Apart from the fact that the $\mathbf{R}_{ij}$ are not univariate scalars, but instead vectors, the formulas are indeed very similar to those of numerator and denominator in the ANOVA F-test. This traditional notation, however, becomes rather cumbersome when leaving the realm of simple designs, whereas the elegant notation allows for a convenient generalization.

3. **Building test statistics.**

   Based on the matrices $\mathbf{H}$ and $\mathbf{G}$, compute one of the nonparametric test statistics that have been developed and validated in the recent research literature. In particular, we recommend use of the nonparametric version of Wilks' Λ, which is defined as

$$\lambda = \frac{\det[(N-a) \cdot \mathbf{G}]}{\det[(N-a) \cdot \mathbf{G} + (a-1) \cdot \mathbf{H}]} .$$

If the matrices $\mathbf{H}$ and $\mathbf{G}$ are singular (not invertible), Wilks' Λ cannot be calculated. This will happen in situations with more endpoints (questions) than survey respondents, and it may also happen in other cases, in particular when the dimension (k) of the response vector is high and there are many ties in the data. A situation of many tied observations is unavoidable when survey outcomes are measured on discrete, ordinal scales with few possible answer categories. In cases where Wilk's Λ cannot be calculated due to singularity, the nonparametric ANOVA-type statistic should be used. It is simply defined as the ratio of the matrix traces, namely

$$T_A = \mathrm{tr}(\mathbf{H})/\mathrm{tr}(\mathbf{G}) .$$

It may appear that the ANOVA-type statistic uses less between-variable information than Wilks' Λ. However, the covariances or correlations between outcomes are considered here, as well. Indeed, they enter into the p-value calculation, which is described in the next and last step.

4. **Finding sampling distributions and *p*-values.**

In order to find valid *p*-values corresponding to ones observations, one needs validated approximations to the true sampling distributions of the test statistics. In particular, the approximations should be designed such that their tail probabilities match the tail probabilities of the true process very well. Correctly approximating the tail probabilities matters so much because typically, statistical inference is concerned with decisions regarding extreme probabilities (less than 0.05 or less than 0.01). For each of the test statistics mentioned above (Wilks' Λ and ANOVA-type), there are basically two well-performing approximations to the sampling distribution.

One of these approximations uses an F-distribution whose degrees of freedom are estimated from the data. This idea goes back to Box [5] and leads to a decent approximation even with moderate to small sample sizes. With small samples, it works particularly well in balanced designs (equal number of respondents per group).

The other approximation uses the permutation distribution of the respective test statistic. The permutation distribution is obtained using these three steps: (1) remove all group labels of all respondents, (2) permute these group labels in all $N!$ possible ways among the respondents, (3) each time calculate the test statistic. The resulting values, each weighted with probability $1/N!$, constitute the permutation distribution. This is actually not even an approximation, but indeed the exact distribution under the null hypothesis of exchangeable random vectors, which is implied by the null hypothesis of equal multivariate distributions. However, in practice, the number $N!$ of possible permutations can get rather large. Thus, one typically resorts to an approximative solution by only considering, say, 1000 or 10,000 randomly chosen permutations. In principle, in the sense of not exceeding the nominal type I error, the permutation distribution approach is expected to work even for rather small sample sizes, but clearly it wouldn't make sense to conduct *any* study if there are too few observations to reach a reasonable power. If there is any doubt whether power may suffice, a good strategy is to perform some simulation experiments before actually starting the data gathering process. In the simulation, synthetic data is generated that should closely resemble the expected real data, along with the effects that the researcher would like to be able to detect (for some ideas on how to generate realistic data, see also Sect. 5.5). If it turns out that the expected effects are not detectable with the study size that can be afforded, it may sometimes be best to pull the emergency brake and prevent resources from being poured into a study that has only small chances to succeed. As a general rule of thumb, we would recommend to have at least seven or eight respondents per subgroup.

5.4 How to Perform These Tests with Statistical Software?

All the steps outlined in Sect. 5.3 can be performed at once using the R-package
npmv (Ellis et al. [9]). The test statistics to be calculated (Wilks' Λ or ANOVA-type)
can be selected by the user. In fact, there are two more test statistics implemented
in npmv whose performance is typically somewhat similar to that of Wilks' Λ (the
Lawley-Hotelling and the Bartlett-Nanda-Pillai tests). By default, p-values based on
both approximations to the sampling distribution are provided (F-distribution and
approximated permutation distribution).

For the simple illustrative example above, the data can be entered into an R data
frame using the following few lines.

```
X1=c(3,2,1,4,2,2,1,4,4,3,4)
X2=c(-0.03,0.64,0.75,-0.80,0.97,0.24,-1.64,-1.79,
    -1.14,-0.29,-0.41)
X3=c(0,0,0,1,1,0,1,0,0,0,0)
group=as.factor(c(rep(1,4),rep(2,3),rep(3,4)))
X=cbind(group,X1,X2,X3)
X=as.data.frame(X)
```

In addition to the numbers given in Table 5.1, we have added a third group of four
three-variate observations. However, please note that this should really be regarded
as a toy example for pure illustrative purposes. In practice, sample sizes of $n_1 = 4$,
$n_2 = 3$, and $n_3 = 4$ would be too small to draw reliable conclusions—even if the
methods presented here provide interpretable answers.

The analysis using npmv can simply be performed with the following two lines
of R-code, assuming that the package has been installed.

```
library(npmv)
nonpartest(X1|X2|X3 ~ group, X, plots=FALSE,
    permtest=FALSE)
```

In the output, no significant differences are revealed for this data set. In other
words, the data have not provided sufficient evidence against the null hypothesis
that the three ($a = 3$) three-variate ($k = 3$) distributions F_1, F_2, and F_3 differ from
each other. Given the small sample sizes, this is not a surprise.

The output also provides the following descriptive summary of the data in terms
of nonparametric relative effects.

```
$releffects
        X1        X2        X3
1  0.44318  0.63636  0.48864
2  0.24242  0.59091  0.69697
3  0.75000  0.29545  0.36364
```

This is to be interpreted as follows. Regarding variable X_2, the estimated marginal
nonparametric relative effect of group 3 equals about 0.3. This means that if a respondent A was randomly chosen from group 3, and if another respondent B was

randomly chosen from among all 11 participants in the study, then, regarding variable X_2, the probability is 0.3 that A's outcome is at least as large as B's. Thus, the nonparametric relative effects provide for an intuitive descriptive interpretation that can be used along with the inferential results.

For data with ties, the wording "at least as large" is interpreted as "at least as large, where equality is only given half weight". This is much easier said with a mathematical formula, namely $P(X_A > X_B) + \frac{1}{2}P(X_A = X_B)$.

5.5 Which *Groups* Differ from Each Other?

Here, we would like to investigate how differences between particular groups may be investigated, in addition to an overall analysis of *all* groups. In classical terms, this is often referred to as a *post-hoc* analysis.

In order to have a data set at hand that is somewhat more realistic in terms of its size and usefulness, let us generate one using the random generator functions in R.

```
set.seed(0)
X1=c(rnorm(15),rnorm(15)+0.5,rnorm(15)+1,rnorm(15)+2)
X2=rnorm(60)
X3=X1+X2+0.5*rnorm(60)
group=as.factor(c(rep(1,15),rep(2,15),rep(3,15),
     rep(4,15)))
X=cbind(group,X1,X2,X3)
X=as.data.frame(X)
```

These data consist of $N = 60$ observation vectors, each of the $a = 4$ groups has $n_i = 15$ respondents, and again the dimension is $k = 3$. The outcomes are generated in such a way that in one of the variables (X_1), the responses tend to take larger values from group 1 to 4. The second variable (X_2) may just be considered as random noise without any information. And, the third endpoint is actually highly dependent on the first two. Such a dependence of responses may occur in practice. Someone who gives high marks for one question may do so for another, related question. In that sense, this simulated data tries to mimic some features that could be encountered in real data.

The analysis can be performed as in the previous example, except that here we don't suppress the graphical output because the number of observations is large enough for it being possibly informative. Also, in this example, we don't suppress the calculation of the permutation test p-value. The only disadvantage of the permutation test is that it takes some seconds to be calculated, but as the output shows, the permutation approach agrees rather well with the corresponding p-values from F-approximations.

```
library(npmv)
nonpartest(X1|X2|X3 ~ group, X, permtest=FALSE)
```

When all four types of tests can be calculated (ANOVA-type, Lawley-Hotelling-type, Bartlett-Nanda-Pillai-type, Wilks' Λ-type), as in this situation, we recommend use of Wilks' Λ-type statistic.

For this example, a highly significant difference between the groups is shown. Between which groups?

The answer can be obtained with another function in the same R package.

```
ssnonpartest(X1|X2|X3 ~ group, X, factors.and.
    variables = TRUE)
```

This function performs a closed testing procedure which reveals all combinations of groups that exhibit a significant difference between them. Ultimately, it shows all pairwise group comparisons that are significant, if there are any. The whole procedure maintains the familywise error rate. That is, all decisions are made simultaneously at the prespecified α, which per default is set to 0.05.

This is reported to the user with the following statement in the output. "All appropriate subsets using factor levels have been checked using a closed multiple testing procedure, which controls the maximum overall type I error rate at alpha= 0.05."

In this data example, significant pairwise differences are found between the following pairs of groups: 1–4, 2–3, 2–4.

Considering how the group differences are introduced in the simulated data in variable X_1, it is not surprising that those groups whose labels are furthest apart from each other (1,4) show a significant difference, but the procedure is also capable of detecting two more significant pairwise differences.

In the function call above, the option `factors.and.variables = TRUE` ensures that group differences are actually evaluated. Since there are fewer variables ($k = 3$) than groups ($a = 4$), if this option is not specified, the algorithm only looks for differences between variables and not between groups.

Differences between variables are considered in the next Section.

5.6 In Which *Endpoints* Do the Groups Differ?

If a global difference is detected, one would usually like to know which of the outcomes have driven this significance. This can be considered a variable selection problem, namely trying to answer the question which variables are most important in distinguishing the groups from one another.

Let us use the same synthetic data already considered in the previous section. The global test indeed revealed a significant difference, so it is natural and legitimate to look for the sources for this difference.

The same code as already presented in the previous section also provides the answers to this quest. The procedure is a modification of closed testing, therefore the wording of the final sentence in the output is somewhat different. Most important is however that also this procedure maintains the familywise error rate at the nominal level.

"All appropriate subsets using response variables have been checked using a multiple testing procedure, which controls the maximum overall type I error rate at alpha= 0.05."

For the simulated data set being analyzed as an example, the procedure has revealed variables X_1 and X_3 as driving the significant differences. Again, the result is not surprising because in the synthetic data, group differences are introduced in variable X_1. Variable X_2 is only noise, so there should be no effect to be detected. And X_3 is highly influenced by X_1, so the differences among groups that are manifest in X_1 also carry over to X_3.

5.7 How to Interpret the Results?

As in most statistical analyses, there are two major components, each requiring careful interpretation. Namely, a descriptive, and an inferential component. In the descriptive part, graphical and numerical summaries of the data are provided. A good visualization should always be the first step of any data analysis, after the typically lengthy data cleaning. In fact, a good visualization may even provide hints that data cleaning needs to continue. Inconsistencies in the data may best be discovered using visual methods. In this chapter, we will not provide detailed advice on how to display data graphically, as most introductory statistics textbooks devote ample space to this topic (see, e.g., [6, 7, 20]).

Part of the descriptive analysis is also the calculation of numerical summary measures. In the context of a nonparametric approach to data analysis, statistics such as means, variances, and standardized mean differences (often referred to as Cohen's d or Hedges' g) are *not* appropriate summaries. They do not make sense for ordinal or highly skewed data, and they have in general no relation to the conclusions from nonparametric inference methods, unless very specific models are assumed (e.g., location shift models).

The most appropriate summary measure is the nonparametric relative effect which is the statistical functional underlying the most important classical nonparametric tests for two and more samples [8, 21, 16, 13, 14]. The same functional is also the basis for the nonparametric multivariate inference methods described in this chapter. For two random variables X and Y, the relative effect is basically the probability that the first one takes a smaller value than the second one. More precisely,

$$p_{XY} = P(X < Y) + \frac{1}{2}P(X = Y).$$

Its empirical analog in case of two samples is the proportion of (X, Y)-pairs from all possible such pairs from the X- and Y-samples where the X-value is smaller than the Y-value. In case of ties (equal values), one needs to add one half of the proportion of pairs where both X and Y take the same value. In other words, the estimated relative effect in two samples is the probability that a randomly chosen observation from the first sample takes a smaller value than a randomly chosen observation from the second sample, in case of ties corrected by one half of the probability of an equal value.

When there are three and more groups of experimental units, one compares the observations in each group with those from a reference sample. Here, a useful reference sample is the combined sample of *all* respondents (from all groups) included in the study. As described in Sect. 5.4, the estimated relative effect of group *i* is the probability that a subject chosen randomly from all subjects in the study yields a smaller value than a subject chosen randomly from all subjects in group *i*. Again, in the presence of tied values, one half of the probability of an equal value is added. "Randomly chosen" here means that each respondent in the respective sets has the same probability of being chosen.

These estimated effects are provided in the output of the R-package npmv (Ellis et al. [9]). They indicate a tendency of observations within a particular group to take larger (or smaller) values, as compared to the other groups. A major advantage of the relative effect is that it does not require metric responses. As long as a "smaller" or "greater" relation can be assessed, the relative effect makes sense. Clearly, larger values of the relative effect in one group indicate that this group tends to larger observed values. A value of $1/2$ for the relative effect can be interpreted as no tendency to larger or smaller values in that particular group.

For the inferential analysis, the relative effects provide an important piece of supplementary information. If the multivariate test reports a significant value, it is instructive to look at the estimated relative effects, in order to be able to interpret the significance.

Regarding the statistical inference itself, we recommend use of Wilks' Λ whenever possible. If this test statistic cannot be calculated, the ANOVA-type statistic should be used. If the *p*-value for an overall multivariate test based on the chosen test statistic is small, there is evidence that in at least one outcome variable there is at least one group whose observations tend to take smaller or larger values than in the other groups, and this evidence is beyond mere chance.

Depending on whether interest focuses on groups or on outcome variables, in a next step, a multiple testing procedure should be performed in order to identify the groups or endpoints that are responsible for the detected significance. The method implemented in the R-package npmv automatically adjusts for the multiplicity of testing. Thus, the probability of making at least one false rejection is controlled throughout the procedure.

Typically, the algorithm yields one or more endpoints or groups responsible for the significant effect. Now, a descriptive reporting of the corresponding nonparametric relative effects aids in interpreting the magnitude and direction of the effect. Finally, a visualization involving the identified groups and endpoints is advisable.

References

1. Bathke, A.C., Harrar, S.W.: Nonparametric methods in multivariate factorial designs for large number of factor levels. J. Stat. Plan. Inference **138**(3), 588–610 (2008)

2. Bathke, A.C., Harrar, S.W.: Rank-based inference for multivariate data in factorial designs. In: Robust Rank-Based and Nonparametric Methods, pp. 121–139. Springer, Berlin (2016)

3. Bathke, A.C., Harrar, S.W., Madden, L.V.: How to compare small multivariate samples using nonparametric tests. Comput. Stat. Data Anal. **52**(11), 4951–4965 (2008)

4. Bathke, A.C., Harrar, S.W., Ahmad, M.R.: Some contributions to the analysis of multivariate data. Biom. J. **51**(2), 285–303 (2009). https://doi.org/10.1002/bimj.200800196

5. Box, G.E., et al.: Some theorems on quadratic forms applied in the study of analysis of variance problems, II. effects of inequality of variance and of correlation between errors in the two-way classification. Ann. Math. Stat. **25**(3), 484–498 (1954)

6. Crawley, M.J.: The R Book. Wiley, Chichester (2007)

7. Dalgaard, P.: Introductory Statistics with R. Springer, Berlin (2008)

8. Deuchler, G.: Über die Methoden der Korrelationsrechnung in der Pädagogik und Psychologie. Zeitschrift für pädagogische Psychologie und experimentelle Pädagogik **15**, 114–31 (1914)

9. Ellis, A.R., Burchett, W.W., Harrar, S.W., Bathke, A.C.: Nonparametric inference for multivariate data: the R package npmv. J. Stat. Softw. **76**(4), 1–18 (2017)

10. Harrar, S.W., Bathke, A.C.: Nonparametric methods for unbalanced multivariate data and many factor levels. J. Multivar. Anal. **99,8**, 1635–1664 (2008)

11. Harrar, S.W., Bathke, A.C.: A nonparametric version of the Bartlett-Nanda-Pillai multivariate test. Asymptotics, approximations, and applications. Am. J. Math. Manag. Sci. **28**(3–4), 309–335 (2008)

12. Harrar, S.W., Bathke, A.C.: A modified two-factor multivariate analysis of variance: asymptotics and small sample approximations (and erratum). Ann. Inst. Stat. Math. **64**(1 and 5), 135–165, 1087 (2012)

13. Kruskal, W.: A nonparametric test for the several sample problem. Ann. Math. Stat. **23**, 525–540 (1952)

14. Kruskal, W.H., Wallis, W.A.: Use of ranks in one-criterion variance analysis. J. Am. Stat. Assoc. **47**(260), 583–621 (1952)

15. Liu, C., Bathke, A.C., Harrar, S.W.: A nonparametric version of Wilks' lambda – asymptotic results and small sample approximations. Stat. Probability Lett. **81**, 1502–1506 (2011)

16. Mann, H.B., Whitney, D.R.: On a test of whether one of two random variables is stochastically larger than the other. Ann. Math. Stat. **18**, 50–60 (1947)

17. O'Brien, P.C.: Procedures for comparing samples with multiple endpoints. Biometrics **40**, 1079–1087 (1984)

18. Pesarin, F., Salmaso, L.: Permutation tests for complex data: theory, applications and software. Wiley, New York (2010)

19. R Core Team: R: A Language and Environment for Statistical Computing. R Foundation for Statistical Computing, Wien, Austria (2017). URL http://www.R-project.org

20. Verzani, J.: Using R for Introductory Statistics. CRC Press, Boca Raton (2014)
21. Wilcoxon, F.: Individual comparisons by ranking methods. Biom. Bull. **1**(6),
    80–83 (1945)